Exploding Disk Cannons, Slimemobiles,

and 32 Other Projects for Saturday Science

Exploding Disk Cannons, Slimemobiles,

and 32 Other Projects for Saturday Science

Neil A. Downie

THE JOHNS HOPKINS UNIVERSITY PRESS
Baltimore

Although every effort has been made to ensure that the information in this book is correct, the publisher can give no assurance to that effect. Neither the author nor the publisher accepts responsibility for loss, damage, injury, and/or death resulting from any act(s) of omission or commission in performing any of the experiments described in the book.

Printed in the United States of America on acid-free paper
9 8 7 6 5 4 3 2 1

The Johns Hopkins University Press
2715 North Charles Street
Baltimore, Maryland 21218-4363
www.press.jhu.edu

Library of Congress Cataloging-in-Publication Data

Downie, N. A. (Neil A.)
Exploding disk cannons, slimemobiles, and 32 other projects for Saturday science / Neil A. Downie.
p. cm.
Includes bibliographical references and index.
ISBN 0-8018-8506-X (hardcover : alk. paper)—ISBN 0-8018-8507-8 (pbk. : alk. paper)
1. Science—Experiments. 2. Science—Popular works. 3. Physics—Experiments. 4. Physics—Study and teaching—United States. I. Title.
Q182.3.D66 2007
507.8—dc22 2006011486

A catalog record for this book is available from the British Library.

The figures that appear at each of the twelve part openers are selected U.S. patent drawings from the years 1905 to 2004.

More Saturday Science . . . If you enjoyed this book and would like more information about Neil Downie's other books or updated details on projects, visit www.saturdayscience.org.

Contents

Preface

> I can't tell you anything . . . you never believe anything unless you've worked it out for yourself.
>
> —Douglas Adams, *The Salmon of Doubt*

This book includes projects and experiments that reflect brand-new ideas or inject new thinking into older ideas. They all illustrate principles and phenomena of science, mainly physics and chemistry. Each project offers details for making one sure-fire version of the experiment, along with hints and suggestions for variations on the project. There are also some detailed science and math analyses. The ideas come from my work in industrial gases and from running a science session at a Saturday morning club for kids in my hometown.

As demonstrations, many of the projects here will fascinate anyone, child or adult. Some of the easier experiments can be easily carried out by kids as young as nine or ten years old. Other projects require greater manual dexterity, so they are suitable only for teenagers and adults. The "science and math" sections contain mostly straightforward algebra but include some calculus and detailed explanations; they are intended for adults and older teenagers.

I have included quite a bit of math under the science and math headings. You can skip these or only glance at them if you like, perhaps on a first reading, because the gist of the experiment and the results are covered in the rest of the text. But I hope you will explore the math sections and perhaps think about a further analysis of the projects when you have some more time.

Math is everywhere, inside everything. Look in detail at things as diverse as a chunk of concrete or a piece of military strategy, and you will find some math. It may be intuitive math rather than pen-and-paper algebra, it may be binary logic rather than numerical analysis—but it will be there.

Take a lump of concrete, for example. Consider the distribution of particle sizes inside it: how does that distribution affect its strength? When you stress a lump of concrete, where does it stretch, and how will it tend to break? Is a lump of concrete stronger with just sand and cement, or is it better to have gravel or pebbles too? And is it better to measure the proportions of aggregates with a particle distribution based on numbers of particles in mass ranges or in diameter ranges (they look quite different)? Good workers repairing an old house with a few bags of cement and aggregate will be thinking about all this while they are doing construction and getting it right intuitively. But if you are building a large bridge or an airport runway, then you probably need to check your intuition with some math.

Doing stuff and thinking about stuff at the same time, it seems to me, are often better than doing either on its own. The question is often asked as to why the flourishing Greek civilizations of the first millennium BC did not go on to develop into a recognizably modern technology-based society thousands of years before the world's first industrial nations arose in Europe. Some commentators have pointed to moral factors such as the Greeks' use of slaves, or to the lack of a strong work ethic such as that of Protestant northern Europe. For me, however, the ancient separation of the theoretical and practical aspects of science offers a more convincing argument. Although the Greeks were capable of great scientific ingenuity, they did not develop technology alongside that science. Perhaps driven by a mentality related to their use of slaves—"slaves do, masters think"—they did not develop scientific applications alongside their scientific experiments. This separation was a continuing theme in science right up to the Renaissance and the early pioneers of modern science, personified by giants like Leonardo da Vinci. Until Leonardo and his contemporaries, nobody bothered to check theory against experiment. Do not, therefore, repeat the mistakes of ancient Greece and the Middle Ages: if, when you read this book, you find something too extraordinary to believe, or something strikes you as a better way than I have suggested, get out of your chair and try it yourself. (And let me know where I went wrong!)

Some projects in this book include elements of danger and so carry safety warnings. Do not, however, be put off doing these projects. If you take reasonable care and observe the specific cautions outlined, they do not involve risks that are any greater than those of everyday life. The safety warnings are there to ensure that you recognize any dangers and behave accordingly, not to discourage you from doing the projects. I don't think I am alone in thinking that kids are often overprotected today, and that this denies them enjoyable and educational experiences that are not in fact unreasonably hazardous. Our overcaution stems partly, perhaps, from our tendency, amplified by the news media, to exaggerate the significance of bad but very rare incidents. You might agree with Nevil Shute, who commented in his book *Slide Rule: The Autobiography of an Engineer,* "To put your life in danger from time to time . . . breeds a saneness in dealing with day-to-day trivialities." I would not, however, go as far as the notorious Nietzsche: "Das Geheimnis, um die gröβte Fruchtbarkeit und den gröβten Genuss vom Dasein einzuernten, heiβt: gefährlich leben!" (The secret of reaping the greatest fruitfulness and the greatest enjoyment from life is to live dangerously!)

I've included a selection of the details involved in these projects. I can't totally agree with Henry Thoreau when he said that "Our life is frittered away by detail . . . simplify, simplify." Sometimes you need a few details to *avoid* frittering. There is a limit to how much can be included, however. I can't include, for example, the ten design variations of the Negative Steam Engine I tried along with the kids at the Saturday club, because it would take too much space. But I hope that each project description includes enough detail to get you something that works to some extent, which you can then improve in your own way. At the end of each project there are hints and suggestions about extending or varying the variables and carrying out further analysis. Obviously, there are few details on these extensions, but I hope that the basic descriptions will get you started.

Finally, I'd like to address those who have read—or might want to read—my previous books. First of all, there are no projects duplicated between this book and my previous Saturday Science books, *Vacuum Bazookas, Electric Rainbow Jelly* and *Ink Sandwiches, Electric Worms.* However, many of the themes illustrated in the projects of my earlier books are followed up in this volume. After the vacuum bazooka and the vacuum railroad in the earlier books, we have in this one the vacuum engine or negative steam engine. On the theme of amplification, the beard amplifier and electrolystor of the previous books are followed here by the piezistor. And after curious vehicles like the duohelicon and the tubal travelator, we have here the ball river bobsled and the slimemobile. So if you like this book, why not read the others? Finally, as I intend to write at least one more book in this genre, I will not say goodbye, but rather au revoir.

I am indebted to many people—more people than I can list here—both for the production of this book and for help or suggestions in carrying out the projects on which it is based. My family deserves special mention. Not only did my wife, Diane, help out with experiments over the kitchen table, but also my two daughters, Helen and Becky, have spent many hours at a computer producing drawings. My work colleagues Ben Inman, Chris Mercer, and Stuart Kerr, and our visiting French chemistry student, Mathilde Pradier, deserve mention for their help with demonstrations, along with my understanding boss at Air Products and Chemicals, Inc., Professor John Irven. Thanks are due to Johns Hopkins University Press editor-in-chief Trevor Lipscombe for his help, for arranging the finishing of the many drawings, and to Peter Strupp and his team at Princeton Editorial Associates. Thanks too to the Royal Institution in London, which provided the venue—the lecture theater famously used by Michael Faraday—for three of my lectures. Finally, of course, a big thank you to the kids at the Saturday Activity Center in Guildford, for whom many of these projects were devised, but who also added some ideas of their own.

Exploding Disk Cannons, Slimemobiles, and 32 Other Projects for Saturday Science

Projectiles

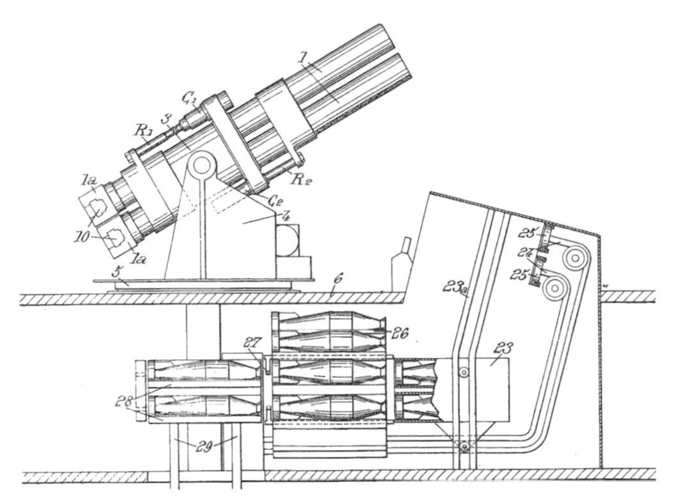

I shall now proceed to an explanation of those instruments which have been invented for defense from danger, and for the purposes of self-preservation; I mean the construction of scorpions, catapultae, and balistae, and their proportions. And first of catapultae and scorpions. Their proportions depend on the length of the arrow which the instrument is to throw.

—Marcus Vitruvius Pollio ("Vitruvius"),
De Architectura, 80–70 BC
(trans. Joseph Gwilt, 1826)*

*Vitruvius's ten books are a treasure trove not only concerning architecture but also on other technologies of the early Roman empire, as in this siege engine description.

Since before the days of epic Greece or ancient Rome, machines that can hurl a large projectile a long distance have been a subject of intense interest because of their use in fighting battles. Today the role of large projectiles in war has been supplanted by jet- and rocket-propelled missiles, and the great guns of the twentieth century are museum pieces. But projectile throwers from trebuchets and ballistas to cannons and howitzers still catch the imagination.

All projectile throwers must in some way store energy that can be released in a short period. This energy must then be applied efficiently to the job of accelerating the projectile. The single project in this section shows how to store an appreciable amount of energy in compressed air and then how to apply that to accelerating a piston projectile in a more or less vertical tube, like a mortar type of cannon. One of the difficulties of many compressed air guns is making a valve that can open both wide and fast. In our project, the ideal valve would have a bore comparable to that of the barrel (35–50 mm, or 1½–2 inches), and because our projectile leaves the barrel within a few tens of milliseconds, the valve should ideally act in a millisecond or two. No ordinary valve can do this. Read on to see how we get around this difficulty.

1 Exploding Disk Cannon

Cannon to the right of them,
Cannon to the left of them,
Cannon in front of them
. . . .
Into the jaws of Death,
Into the mouth of Hell
Rode the six hundred.

—Alfred, Lord Tennyson, *The Charge of the Light Brigade*

Most things we use are designed not to break. It is wasteful if something breaks and has to be replaced all the time: it is usually much better to make it a little stronger and longer lasting. However, while devices that are specifically intended to break are exceptional, disposables do have a long history. The spears used by the legions of ancient Rome, for example, had a small barbed tip at the end of a short, thin metal shaft, joined onto the main spear shaft, which was wood. One problem with ancient battles was that projectiles were often reused. If you threw some heavy stones at the enemy and missed, you were quite likely to be targeted by your enemy using the very stone you had just thrown. The same applied to spears, until the Romans came along. The Roman legionnaire's spear was perfectly strong and deadly when it hit its target squarely. If the spear missed its barbarian target, however, it would generally strike the ground obliquely and be bent. A bent spear could not be turned against its original owner, at least not in the thick of battle. After a battle, many of the spears would be recovered by the victorious army and hammered back into shape for reuse. But during the battle they were nonreusable.

A couple of millennia later, engineers devised a nonreusable way of protecting pressurized vessels in gas systems from excess pressure by means of a small, thin disk of corrosion-resistant metal. Look closely at a compressed gas cylinder, or at a large tank of industrial liquefied gas such as liquid nitrogen or liquid oxygen, and you will often find an apparently functionless fitting or dead-end pipework branch. There may be a device called a "relief valve" fitted in this branch. If the vessel or tank is accidentally subjected to excess pressure, it will simply release a small amount of gas via a spring-loaded valve—a valve similar to the small weight-loaded valve on the top of a pressure cooker. However, if that valve is also subject to accidental failure, or if there is no relief valve, then the backstop, the ultimate protection against the explosive destruction of the pressure vessel, is most often a bursting disk. At a critical pressure, the disk will rupture completely, opening a hole with a large bore to vent all the gas in the tank to avoid a serious failure.

Our project here also relies on the catastrophic failure of a bursting disk, but one in which the failure is initiated by a pin.

What You Need

- ❏ Empty soda bottle with unbent cap
- ❏ Sticky tape
- ❏ Tubing of 5 mm ($^{3}/_{16}$ inch) outside diameter
- ❏ Short piece of steel tube of 4 mm ($^{1}/_{8}$ inch) outside diameter (e.g., a car's brake pipe)
- ❏ Hot-melt glue
- ❏ Pressure gauge
- ❏ Air pump
- ❏ Small drain pipe of 35 mm ($1\,^{1}/_{2}$ inches) diameter, at least 0.5 m (18 inches) and preferably 1 m (3 feet) long
- ❏ Softwood cylinder to fit pipe
- ❏ Expanded Styrofoam for nose of wood cylinder
- ❏ Thumbtack
- ❏ Washer

What You Do

The exploding disk cannon comprises a source of compressed-air energy, a soda bottle fitted with a bursting disk; a trigger device—a drawing pin—to burst the disk at the right moment; a launching tube; and a projectile. The bursting of the disk results in a blast of compressed air that can carry the projectile hundreds of feet.

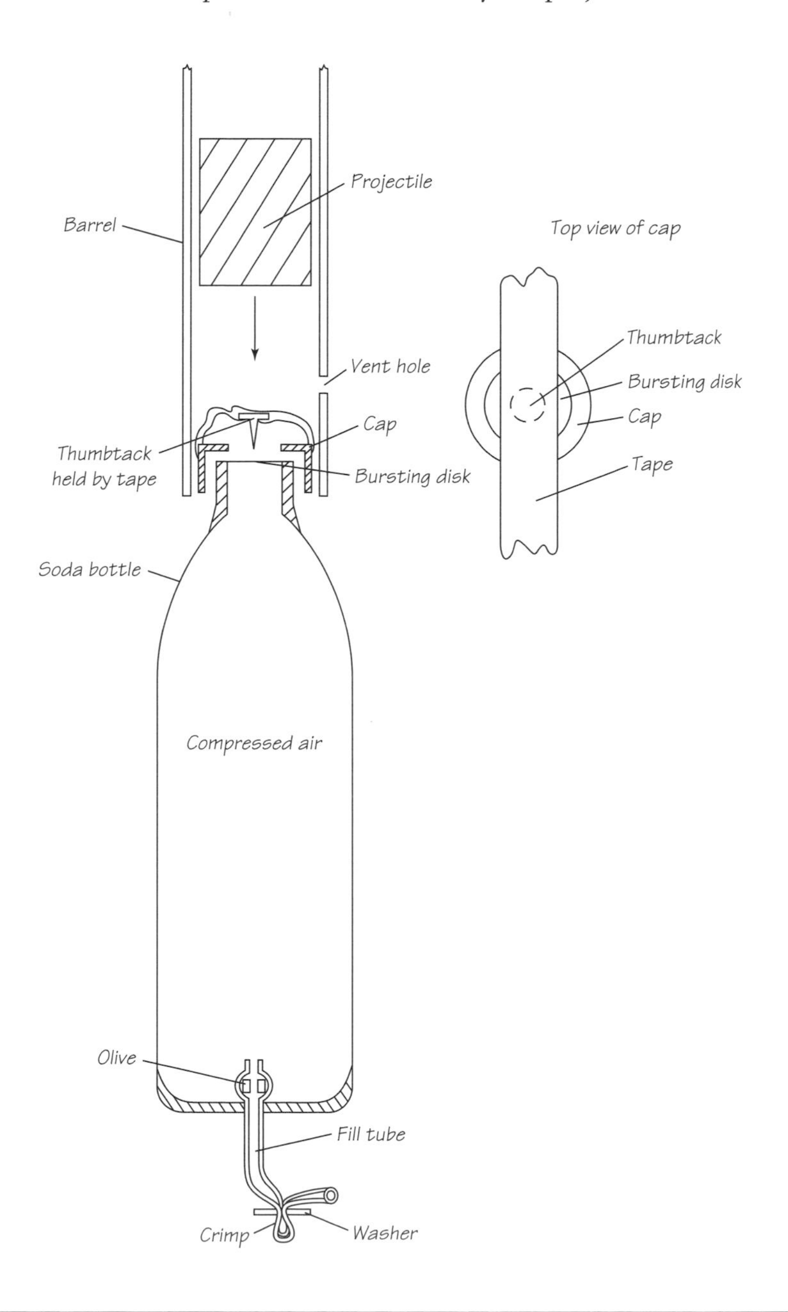

The projectile must fit in the launch tube snugly but freely, easily capable of dropping through, propelled only by the force of gravity. Round off its edges and use a light softwood, or a very hard balsa wood, to minimize any damage caused by impacts or ricochets. I tried using a champagne cork, figuring that it would cause even less damage should it hit anything, and indeed corks do work quite well. They are just a bit too light, however, and need some modeling clay added to give them enough weight to trigger the bursting disk reliably. (There are other problems: You have to wait quite a while for a cork to stabilize in size after drawing it out of a bottle, as it will be tightly compressed. And you have to sandpaper the cork to fit it neatly in the barrel. Finally, you have to find time to drink the champagne!)

You prepare the soda bottle by drilling a small hole, a neat fit for the 5-mm tubing, in the center of the base. Fit the length of tubing at the end with an insert, an "olive," made from a 4-mm slice of automobile brake pipe or other steel tube. Fit the tubing from the inside of the bottle, threading it first through the neck of the bottle and then the base, perhaps with the aid of a piece of coat-hanger wire inserted through the base hole as a guide. Secure the tubing within the base hole by pulling it firmly. The olive, by bellying out the plastic tubing slightly, will ensure that, once the bottle is pressurized, the gas inside will create a more or less air-tight fit. Applying a little silicone sealant before insertion, or a little superglue (cyanoacrylate) before or after fitting, may help effect a complete seal. Secure the tubing from the outside with some hot-melt glue to prevent the tube from being bent too close to the bottle, which might cause a leak.

WARNING

Don't try inflating a soda bottle without a bursting disk. Also don't try to exceed 3 or 4 barg (45–60 psig) of pressure on the soda bottle, and don't use too large a bottle. A ½- or 1-liter (16.9 or 33.8 fluid ounces) bottle is sufficient; a larger bottle will not make a larger bang or provide a noticeably faster projectile. Soda bottles are blow molded out of PET polyethylene terepthallate, a polyester polymer; see Crawford, *Plastics Engineering*), which is quite strong, particularly because it is biaxially oriented and stretched during the molding process, but not unbreakable. If you're unlucky enough to break the soda bottle itself, tiny but still unpleasant or even dangerous fragments of plastic might hit you. The smaller soda bottles will withstand higher pressures than the larger ones.

Drill out the soda bottle's cap with a 13- or 15-mm (½- or ⅝-inch) drill. (This is usually the largest size that will fit readily available small handheld drills.) Close the top of the bottle with a small square of sticky tape and replace the bottle cap.

The trickiest part of the exploding disk cannon is ensuring that the "firing pin" (the thumbtack) is in just the right place to puncture the bursting disk correctly, and is retained and not fired out of the cannon like the piece of wood. You could glue the thumbtack to the bottom of the projectile, but this would make the cannon more hazardous. You should be careful to aim so as to avoid hitting anyone, of course, but if the projectile did by mischance hit someone, then the presence of a thumbtack on it would be clearly undesirable, even if the tack were on the back. The projectile tends to tumble in the air or to bounce off the ceiling, and it might hit someone pin first. Someone might also pick up or catch the projectile and be injured by the pin. That is why I suggest attaching the thumbtack to the tape in the manner described.

You hold the barrel tight onto the soda bottle's neck, but the soda bottle must form a reasonably airtight seal—no gaps bigger than 1 mm (under $^1/_{16}$ inch) —with the barrel. If necessary, glue a ring of plastic 6 mm ($^1/_4$ inch) high inside the bottom of the barrel to achieve this seal. Finally, another feature needs some explanation: the small side vent hole near the bottom of the barrel. Its raison d'être is to allow the air to flow reasonably freely from beneath the projectile as it drops down the tube. Otherwise the projectile will drop too slowly to burst the disk. The vent hole should be about 5 to 6 mm ($^3/_{16}$ to $^1/_4$ inch) in diameter.

Now you are ready to launch. Aim the barrel nearly vertically upward at a harmless target or preferably in the open air. An alfresco launch site such as your backyard may be best until you are used to how the device fires. Pump the soda bottle up to a low pressure—start with just 1 bar gauge (15 psi). Seal off the bottle by crimping the tube with the aid of a washer before disconnecting the pump. Push the soda bottle onto the bottom of the barrel, resting the bottle on a tabletop or holding it firmly against the barrel. Now drop the projectile down the barrel. It should drop, slowed only slightly by friction and the air forced out of the vent hole. A second later it will strike the thumbtack, and then the bursting disk will shatter and, with a muffled bang, the projectile should shoot out upward.

WARNING

Always treat the exploding disk cannon with respect. The projectile may launch at well over 100 mph. I strongly advise that you make the barrel at least 3 feet long, and that you operate the cannon on your own. If you are holding the bottom of the cannon and the soda bottle and also inserting the projectile into the top of the barrel, then you can't get your eyes or any other parts of your body near the projectile path if it must travel through a 3-foot barrel. One of the most frequent emergencies at our local hospital trauma unit—and I would suppose at lots of others—is eye or facial trauma from corks ejected from champagne bottles at up to 100 mph. Don't add to these statistics.

How It Works

This project relies on the "potential energy" stored within the "spring" of a compressed gas. (I use the quotation marks because, at a microscopic level, the stored energy of a gas is in its constituent gas molecules, in their kinetic energy.) The gas is compressed to a smaller volume within the soda bottle, storing an amount of energy E approximately proportional to the volume V of the bottle and the absolute pressure P to which it is raised, or, to put it mathematically,

$$E \sim PV.$$

You can understand this relationship by thinking about a piston compressing a gas within a cylinder. Forcing the piston of area A down against the gas pressure P will involve pushing against a force F, where $F = PA$. Pushing that piston through distance L means doing work—storing energy—equal to $E = FL$. But FL is just PAL, and AL = change in volume V, so $E \sim PV$. A detailed look at the

WARNING

The exploding disk cannon has enough energy to knock holes through Styrofoam ceiling tiles, and it will certainly break unshielded filament lamps and fluorescent tubes.

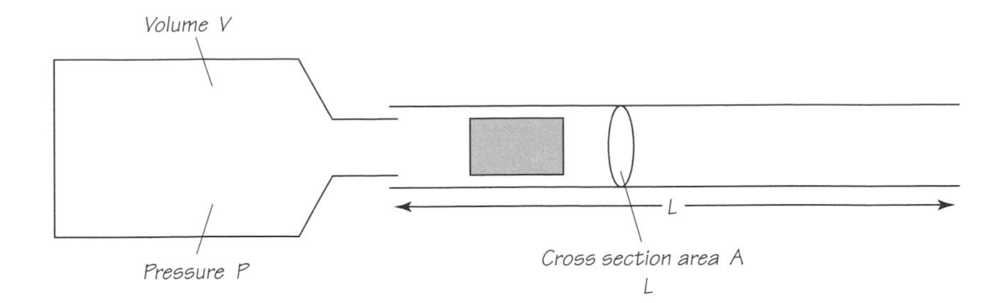

mathematics (see below) shows that the energy stored is slightly larger than this by a logarithmic factor.

You might expect that if you make a small hole in a plastic film with air pressure behind it, then all that will happen is that air will simply leak out of the hole. This certainly can happen. Attach some sticky tape firmly on an inflated rubber balloon and then try to pop the balloon with a pin through the tape. Instead of bursting, the balloon will remain intact and slowly deflate through the pinhole. More commonly, of course, attacking a balloon with a pin results in a bang and the complete destruction of the balloon, and this is the result we need to make the exploding disk cannon work. Similarly, you may be unlucky, and use some type of sticky tape that will absorb the thumbtack impact and survive with only a pinhole. Most types of tape that I've tried worked quite well, however: they shatter completely, leaving the complete bore of the cap free for air to flow out and propel the projectile.

Explore the effects of different charging pressures. What is the lowest pressure at which the disk actually bursts when the pin pierces it? Try different kinds of tape as your bursting disk. Some will inflate like a balloon and won't stand the charging pressure. Try leaning the tube over away from the vertical: at some angle to the vertical, it will stop working because the projectile will not push the pin hard enough.

THE SCIENCE AND THE MATH

The energy in a volume of compressed gas can be calculated by considering the work done by that gas in expanding against a piston. Suppose, for example, that a piston has area A, the original volume is V, and that the gas at pressure P_i is expanded until its pressure reaches P_f, which might be, for example, the pressure of the atmosphere:

$$\text{Energy } E_{comp} = \text{force} \times \text{distance} = \int_{V_i/A}^{V_f/A} PAdx$$

The perfect gas equation, which approximately describes the behavior of air and airlike gases under most ordinary conditions, is:

$$PV = nRT$$

where P is pressure, V is volume, n is the number of moles (the number of gas molecules divided by Avogadro's constant) of gas, R is the gas constant, and T is the absolute temperature.

Using the perfect gas equation, we have

$$P = nRT/V.$$

$$\text{So } E_{comp} = \int_{V_i}^{V_f} nRT\, dV$$

since $Adx = dV$.

The perfect gas equation also gives us the value of V_f in terms of P_f:

$$V_f = nRT/P_f$$

and we end up with

$$E_{comp} = P_i V_i \log_e(P_i/P_f).$$

Given an energy, and a projectile mass m, we can estimate the muzzle speed of our projectile, assuming that almost all the gas energy E_{comp} is converted into kinetic energy $\frac{1}{2} mv^2$:

$$v_{muzzle} = \sqrt{(2E_{comp}/m)}.$$

Substitute typical values for the various parameters, and you will see that a light wood projectile weighing about 20 g could accelerate up to an impressive 200 ms^{-1} (450 mph), if all the energy in the air was converted efficiently. This is a performance not quite equivalent to that of the gunpowder-driven cannons of yesteryear, which were able to fire a projectile at roughly the speed of sound (Mach 1, or 340 ms^{-1}).

Perhaps fortunately, however, the energy of the gas in the soda bottle is not converted to projectile energy with anything like this efficiency. Perhaps the first source of inefficiency to consider is the effective final pressure P_f. Unless the barrel has just the correct volume, then the projectile will either leave it with some air-pressure energy still unused, or it will more or less fully expand the compressed air and will actually slow down before its leaves the barrel. In a typical design with about 1 m (3 feet) or so of 32-mm (1¼-inch) pipe, the pipe's internal volume will be only 0.6 or 0.7 liters, barely more than the volume of the soda bottle. As a result, we can probably assume $P_f \sim P_i/2$, so that instead of $\log_e 5 = 1.61$, we have only $\log_e 2 = 0.69$, leaving only 43 percent of the theoretical energy.

A second source of inefficiency is the sudden expansion of the air: this cannot possibly be efficient isothermal expansion. It must instead be a less efficient adiabatic expansion. Adiabatic expansion occurs when, instead of pulling in some thermal energy from the walls of the barrel and staying at a constant temperature, the air cools down during the expansion. If heat came in from the outside during the expansion of the air, it would follow Boyle's law:

$$P_i V_i = P_f V_f.$$

Instead, it follows the adiabatic equivalent of Boyle's law:

$$P_i V_i^{\gamma} = P_f V_f^{\gamma}$$

where γ is the ratio of the gas-specific heats at constant pressure to constant volume. This is about 7/5 (1.4) for air, which is a mixture mostly made up of the diatomic gases oxygen and nitrogen.

The work W done on the projectile by the air is simply:

$$W = \int P dV.$$

In the case of adiabatic expansion,

$$PV^{\gamma} = k$$

where k is a constant.

$$\text{So } P = k/V^{\gamma}$$

which means the work done is

$$W = \int_{V_i}^{V_f} k/V^{\gamma}\, dV$$

or equivalently,

$$W = \Big[-(kV^{\gamma+1})/\gamma\Big]_{V_i}^{V_f}.$$

A bigger source of inefficiency occurs because all the air from the soda bottle must pass the burst disk, whose diameter—depending upon your drilling of the bottle cap—is just 12 or 15 mm or so. If we assume that the air passes this choke point at approximately the speed of sound, then we might expect the maximum average air velocity in the muzzle to be given at least approximately by the ratio of the diameters squared, that is,

$$v_{max}/D_{muzzle}^2 = v_{sound}/D_{cap}^2$$

where v_{max} is the achieved maximum speed, v_{sound} is the speed of sound, and D_{muzzle} and D_{cap} are the diameters of muzzle and cap hole. This would suggest a much more plausible maximum muzzle velocity v_{max} of only 75 ms^{-1}, or around 150 mph—although still impressively fast. Such a projectile, if streamlined, would travel an equally impressive 600 m or nearly half a mile, if launched at a little less than 45 degrees.

As it happens, 45 degrees is the angle of maximum range for a projectile in a vacuum. The angle of maximum range is less than 45 degrees in the presence of air friction. The maximum range is on the order of v^2/g, which you can understand as follows:

In the parametric equations for the flight of a projectile launched at speed v and angle θ to the vertical:

Vertical velocity v_y at time t is given by

$v_y = v \cos\theta$ at launch and, subsequently, $v_y = v\cos\theta - gt$ as it is accelerated downward by gravity g. By symmetry, the vertical velocity at landing will be $-v\cos\theta$. So we can see that, at landing:

$$gt = 2v\cos\theta$$

and so

$$t = \text{flight time} = 2v\cos\theta/g.$$

The horizontal velocity v_x is given by

$$v_x = v \sin \theta.$$

Hence the horizontal distance traveled is $L_{max} = v\sin\theta \cdot t = v\sin\theta \cdot 2v\cos\theta/g = v^2/g$ if we assume a 45-degree launch (it is left as an exercise for the reader to prove that a 45-degree launch angle provides the optimum range).

So $L_{max} \sim 75^2/9.8 \sim 600$ m.

Bursting Disks: Why Do They Burst?

Why does the cellophane tape shatter, rather than just leak, when it is pierced by a pin? The answers lie in the way energy is released from the surface tension in the tape during the creation of a split, and in the way energy is lost during the creation of the split. If we assume that it takes an energy of S per unit length to cut through the tape, then, for a length L of split, the energy lost E_l is given by:

$$E_l = SL.$$

Meanwhile, energy is being released by the existence of the split, from an area of tape roughly given by L^2. If the surface tension is σ the energy released E_r is given by:

$$E_r = \sigma L^2.$$

Now if $E_r > E_l$, then the crack will grow, driven by the gain in energy which follows further crack creation. So, if $\sigma L^2 > SL$, the crack will grow, that is, the crack growth criterion is:

$L_{min} > S/\sigma$.

In other words, there is a minimum length of crack L_{min}. Cracks bigger than L_{min} will grow catastrophically and destroy the tape completely; smaller cracks will simply sit there doing nothing. A naked balloon has a sufficiently low value of S, the energy needed to create a unit of cracking, so that L_{min} is small, and even a small pinhole will completely shatter the balloon. With tape affixed to it, the balloon has a higher value of S, and so a small pinhole in the tape on the balloon just sits there. However, the surface tension σ in the tape of the bursting disk is much higher than with the balloon, because we are using a much higher pressure; therefore L_{min} is small, and, as in the naked balloon case, even small cracks like a pinhole will grow almost instantaneously and shatter the whole disk.

Optimum Lengths for a Cannon Barrel

To choose the optimum length for the barrel, many other factors of practicality enter into the discussion. You may be surprised to learn, however, that the maximum length of the barrel is ultimately limited by straightforward calculations. If we assume isothermal expansion of the compressed air, then the calculation is simple indeed. We use Boyle's law, and we require our barrel to add speed to the projectile. No speed is added once the air has expanded to atmospheric pressure P_a:

$$P_a(AL_{max} + V) = P_i V$$

where A is the area of the barrel cross section, L_{max} the maximum usable length, V the soda bottle volume, and P_i the charging pressure.

This gives $L_{max} = (P - P_a)V/(AP_a)$ or

$$L_{max} = V/A\ (P/P_a - 1).$$

This relationship takes an extremely simple form if we measure the pressure in barg (gauge pressure atmospheres relative to 1 atmosphere pressure) rather than bara or any other absolute pressure measure. With P in barg, then,

$$L_{max} = PV/A.$$

However, we know that the expansion in the barrel is largely adiabatic. This means that once again we must use the adiabatic form of Boyle's law:

$$P_a(AL_{max} + V)^\gamma = PV^\gamma$$

which leads to

$L_{max} = V/A\ ([P/P_a]^{1/\gamma} - 1)$, which is the maximum barrel length equation.

Now let's run some numbers. A 0.5-liter soda bottle with a 4-barg charging pressure on a 30-mm caliber barrel leads to a maximum barrel length L_{max} of 3 m. The factor $(5 - 1) = 4$ for the isothermal case changes to $(5^{5/7} - 1) = (3.16 - 1) = 2.16$ in the isothermal case, leading to a barrel length of L_{max} half as big, just 1.6 m (5 feet).

And Finally . . . Why Bullets Have Skirts

Although the muzzle velocity of the exploding disk cannon is impressively high, the projectiles will not travel a particularly long way if lobbed into the air, despite the calculations above. This is because a cylindrical projectile has very high drag, especially if it is tumbling. If you have access to a wood lathe, try making a streamlined projectile with a round nose, fitting a long, tapering tail

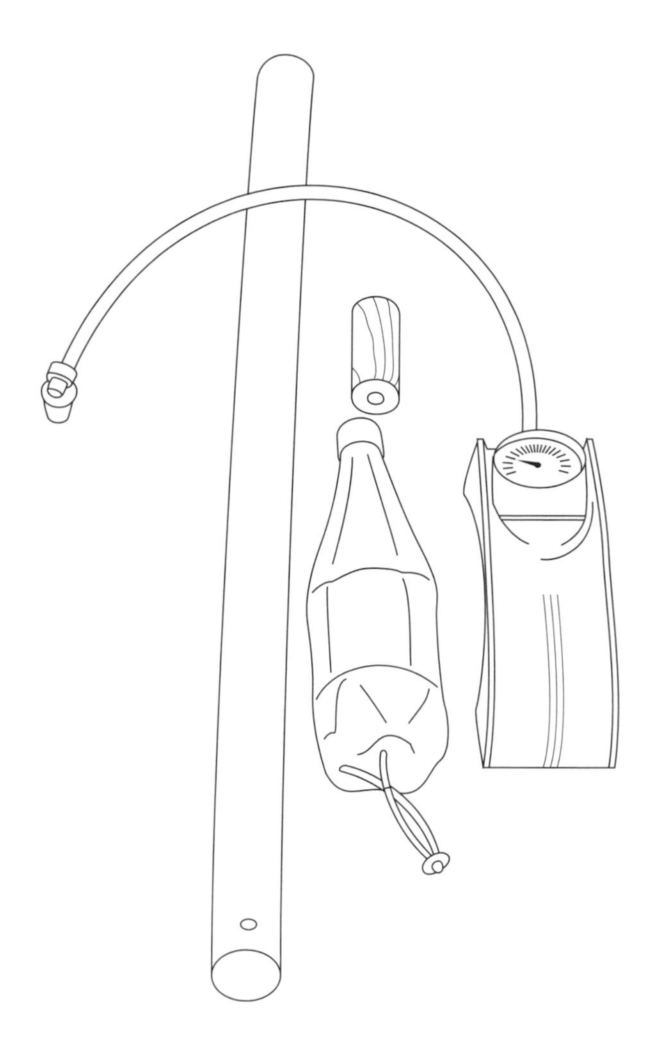

with fins to stabilize it. Alternatively, can you think of any way to make the projectile spin, stabilizing its flight?

The bores of most military cannons have been "rifled," that is, machined with helical grooves, since the mid-nineteenth century. Indeed, the name of a portable weapon incorporating rifling—the rifle—reflects this design. The bullet for a rifle has a skirt or ring that fits tightly within the bore, gripping in the rifling and causing the projectile to spin as it goes up the barrel. The projectile is given a twist of typically only a fraction of a turn in the length of the barrel. This is enough, however, to ensure that it will then continue spinning on the axis of its flight, the gyroscopic action thus stabilizing its direction. This stability is precisely the reason why a quarterback throws a football with a nice, tight spiral.

Could you use a bigger-diameter bursting disk by boring the soda bottle cap to 15 or 17 mm? If you make the hole too big, the tape won't stand the 4-bar charging pressure. You could double up the tape or use a stronger tape to counteract this. But does a wider bore given to the air released from the soda bottle significantly help to increase the power of the device? Can you estimate the flow from the bottle with different sizes of bursting disks?

REFERENCE

Crawford, R. J. *Plastics Engineering*. 2nd ed. Oxford: Pergamon Press, 1989.

Designer Demolition

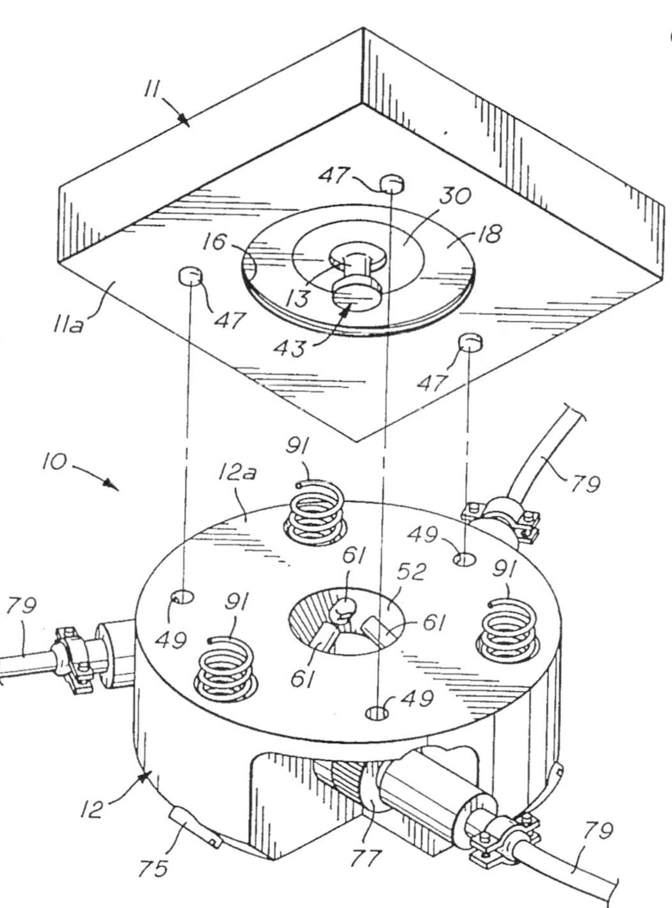

Cogito ergo boom.

—Susan Sontag, *Styles of Radical Will*

Demolition is a source of fascination. No large structure can be dynamited without a surrounding fence to keep back the crowd that inevitably turns up to watch. I am fascinated by demolition myself—particularly by the curious, almost plastic, behavior of apparently solid materials during the few seconds that a building takes to fall. You might think that explosives are the bad guys of the chemical industry, and it is certainly true that modern explosives bear a heavy responsibility for the destruction of warfare today. But they have also allowed large-scale mining to be undertaken with a minimum of cost and, even more important, a much lower toll in the injury and death of miners. In demolition too, explosives have prevented many deaths and injuries. The destruction of buildings piece by piece is a dangerous game, particularly with large buildings. In the bad old days, crews would have to climb to the top and destroy each subsequent floor of even the largest constructions.

These days, however, explosives allow much more precisely designed demolition that minimizes "collateral" damage to adjacent buildings. Large buildings, for example, can be "imploded." This occurs when a building is made to collapse within its own walls. Demolition implosion is achieved through an initial collapse of the innermost support columns in a building, followed seconds later by explosions in the outer columns: with correct design, the outer parts of the building will collapse into the partial hole caused by the initial central collapse. The key advantage of demolition by explosives, however, is that it avoids putting demolition workers in danger, keeping them a long distance from falling masonry.

Explosives are not essential to the idea of remote demolition. An ingenious technique, before explosives became common, was to replace part of a building's base with wood struts and then to light a fire around the base. Once the fire burned through the struts, the building would collapse along a line toward the struts. This technique is still occasionally practiced with brick chimney stacks dating from the nineteenth-century industrial era. These old stacks, often unsafe because of the corrosion of the cement holding them together, are difficult subjects for demolition. They are too tall to attack with a conventional crane and too dangerous for workers to climb. For these reasons, however, they are ideal subjects for "bonfire demolition," and they can accelerate their own destruction in an ingenious way: the draft from hot gases going up the stack will intensify the heat of the bonfire, speeding up the process of burning through the supporting struts.

Here we explore two other possibilities for demolition without the use of explosives. The first project aims to pop party balloons electrically: surprisingly, perhaps, this bit of fun turns out to contain quite a bit of interesting science and math. The second project resembles the famous exploding bolts beloved of NASA space rockets, providing a similar functionality without resorting to pyrotechnics.

2 Balloon Detonator

Necessarily we are all fond of murders, scandals, swindles, robberies, explosions, collisions, and all such things.

—Mark Twain, "Italian without a Master"

The art director will want the girl prettier, want her to wear sexier clothing, want the explosions bigger, want the spaceship bright red instead of dull grey, want the hero tall and handsome instead of short and ugly.

—Roger MacBride Allen, sci-fi author

A galaxy of literary greats have confirmed that explosions are important, in print if not in reality. There are few Hollywood blockbuster movies without some explosions, and the more expensive movies seem to have the most—more bang for more bucks, you might say.

At the ends of parties with balloons, it is almost compulsory that a few get popped. At some events, balloon popping is even a planned activity. Decorators for large events sometimes employ pneumatic balloon exploders (the reference at the end provides one of several suppliers). I have not used these devices, but they typically involve a small plastic piston with a point poking into a balloon to explode it, driven by pressure from a regulator on a gas cylinder (often the helium used to inflate the balloons). These gadgets work in an obvious way, but they are expensive. The electric detonator we develop here is both more ingenious and simpler, as well as being more interesting to analyze.

What You Need

- ❑ Balloons
- ❑ Nickel-chromium (NiCr) resistance wire, 28–31 AWG (30–34 SWG or 0.31–0.24 mm diameter)
- ❑ Nickel-cadmium (NiCd) batteries
- ❑ Connecting wire
- ❑ Tape
- ❑ Standard electrical switch with momentary "on" action (e.g., push-button type); must be a high-current type, capable of at least 5A or 10A
- ❑ Optional: active flux-cored solder (the type sold for joining aluminum), plus soldering iron

For testing helium versus air balloons

- ❑ Helium gas
- ❑ Sound (decibel or dB) meter

Optional: to fill balloons to same pressure

- ❑ Tubing and wood to make a U-tube manometer

Optional: for balloon building demolition

- ❑ Large number of child's wood bricks
- ❑ Two plywood boards, 6–12 mm thick, 300 mm or more square
- ❑ String

What You Do

Our balloon detonator uses a section of NiCr wire, heated electrically, which is taped to the side of an inflated balloon. Using conventional demolition equipment language, the "exploder" is the device that activates an explosion, in our case the switch and batteries that trigger the device. The "detonator" is the device that initiates the main charge, in this case the balloon, and so our heated wire is arguably the detonator.

First you need to wire the battery and switch for your exploder, as shown in the diagram. Then you need to connect long wires to a short (about 2 cm [1 inch]) length of NiCr resistance wire. You can simply twist the bared ends of the copper wires with the NiCr wire. Alternatively, you can make a more reliable

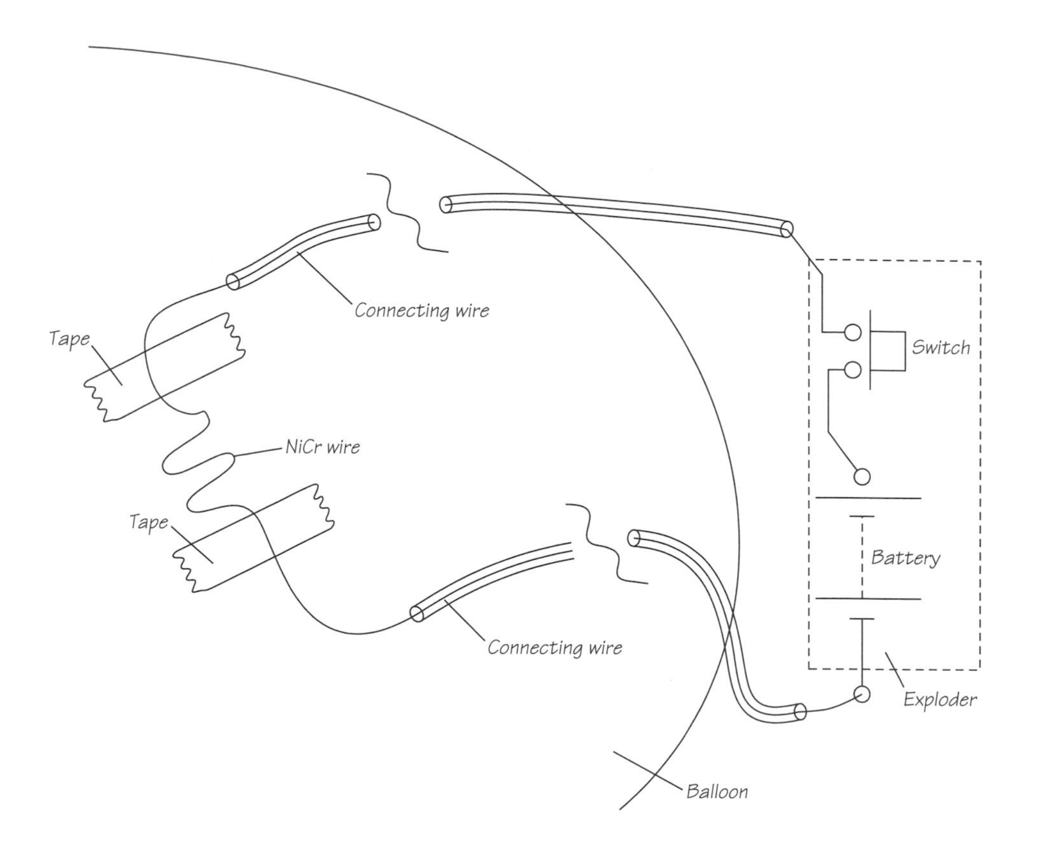

connection with a soldering iron, using a solder with a powerfully active flux core, like that sold for joining aluminum wires.

Take care to tape the wire on one side, over the ends of the wires and the copper lead wires, but with the NiCr wire bent slightly and held by the tape in such a way that the middle part touches the balloon skin without your taping over the middle part. If you tape over the back of the NiCr wire, it will reinforce the balloon skin and stop it from exploding. Inflate the balloons fully for maximum effect from the higher pressure and higher volume. Now press the exploder button. You shouldn't have to wait long for the bang!

Try changing the exploder voltage. With more batteries, more current will flow, and the wire will heat more quickly, giving a more instantaneous bang. But how many batteries is it worth adding?

Try exploding a helium balloon, with a decibel meter listening for the peak sound intensity. These meters normally take a reading of sound intensity averaged over 1 or 10 seconds, but they also usually have a function that will hold the maximum intensity registered. Compare the maximum sound intensity reading from a helium balloon to the reading from an air-filled balloon. Now ask the people around you which explosion they thought was louder—you may be surprised at what you find.

To be fair with this test, you need to use balloons of the same size and type, and filled to be the same size or at the same pressure. You could assemble a U-tube manometer about 1.5 m (4½ feet) high to measure the pressure, connected via a T-piece to the balloon, to ensure this equivalence. All you need is a U-shape of tubing taped to a vertical piece of wood, half-filled with colored water. You can obtain an increased pressure, at least doubled, by using two balloons, one inside the other, and inflating the inner one. Large balloons doubled this way make quite an impressive bang.

Compressed Balloons for Demolition

The exploding balloon can be used for demolishing "buildings" made from stacked wood bricks, by sandwiching a balloon between two wood boards. The boards should be held together by strings, with a large, strong balloon pushing them apart. When the balloon explodes, the boards will collapse, bringing the building down.

To set up this demonstration, first find two rigid plywood boards, a foot (300 mm) or two across. Now drill four 3-mm (⅛-inch) holes in them near the

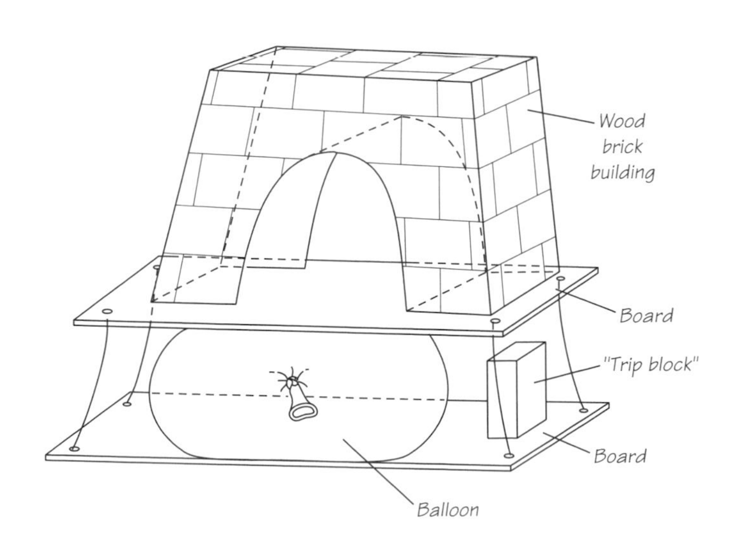

corners, making sure that the holes in the top board are a bit further in from the edges so they don't quite line up with those in the bottom board. This will make the final assembly sit more steadily. Now take four lengths of string and tie knots in them, so that the boards can be held apart by at most 75–100 mm (3 or 4 inches), although this distance will depend upon the size of balloon you use. Make sure that all the strings are in tension as you pull the boards apart to test the strings.

Now you are ready to install the balloon. Place the balloon between the boards, inflate it until it is under high pressure, pushing the boards apart with as much force as possible. Now tape on a balloon detonator and start building with the blocks. The bigger the pile of blocks, the better the demolition effect. One additional trick: try adding a "trip block" to one side of the compressed balloon assembly. Otherwise you may find that the building may simply move vertically downward until the boards meet, without causing collapse of the building.

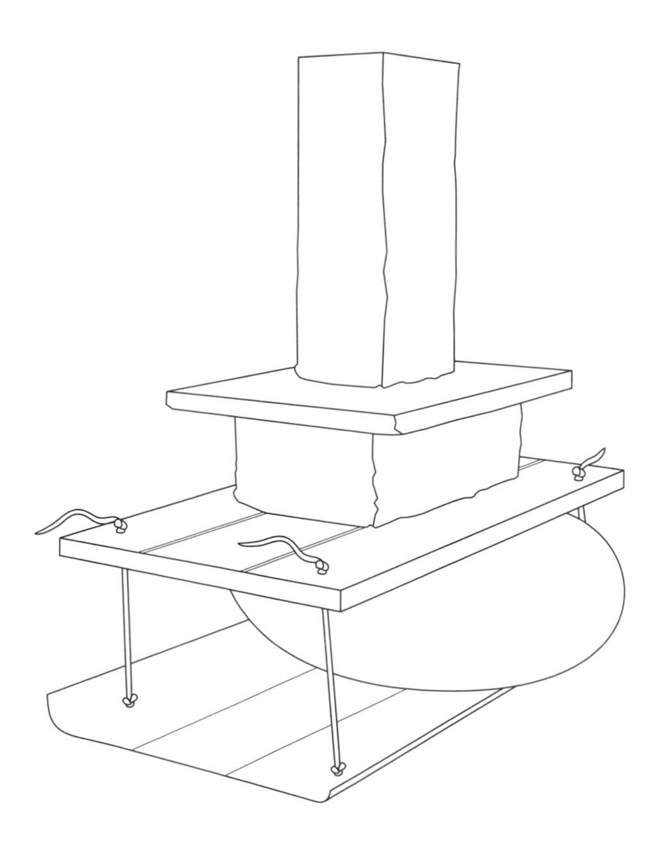

If you puncture a balloon, it explodes rather than simply leaks because of the catastrophic spread of cracks across its surface. Review the analysis of the exploding disk cannon in the previous section; if the crack exceeds a critical length L_{min}, under a certain stress or surface tension σ:

$$L_{min} > S/\sigma$$

where S is the energy per unit length needed to cut the rubber wall of the balloon.

In this case, the crack that we start has a length probably only equal to a fraction of the heated wire's length, but this is still plenty long enough to allow the balloon to burst with a bang. We know that L_{min} is less than the diameter of a pin, since otherwise we wouldn't be able to burst balloons with a pin. With a piece of tape reinforcing the balloon behind the hot wire, this will no longer be the case—which is why if you put tape behind the detonator wire you won't get a bang, you'll get a whimper.

We can calculate the time before the balloon explodes as follows: the wire will puncture the balloon when its temperature exceeds the temperature at which the balloon's latex rubber loses strength. Calling that temperature T_{ex} and room temperature T_r, the heat energy E required for the wire to reach that temperature is approximately

$$E = M_w C(T_{ex} - T_r)$$

where M_w is the wire mass and C is the specific heat capacity of the wire. But the power P applied to the wire is given by the electrical power formula:

$$P = VI$$

where V is the voltage applied and I is the current flowing. Using Ohm's law, $V = IR$, we have

$$P = V^2/R$$

where R is the resistance of the wire. Since power is simply energy per unit time, the amount of time taken to accumulate the necessary energy is given by:

$$P/t = E \text{, so } t = E/P.$$

The time t taken to explode is thus given by the balloon detonator equation:

$$t = M_w CR(T_{ex} - T_r)/V^2,$$

which clearly shows that the exploding time will decrease rapidly as the voltage applied is raised. (Double the voltage and the exploding time will be slashed by a factor of four.)

A More Precise Balloon Detonator Equation

This fairly simple equation ignores, of course, the fact that there are heat losses from the wire, especially if the wire heats up toward the T_{ex} rather slowly. Heat power Q will be lost from the wire approximately according to Newton's law of cooling, which says that the rate of loss of heat depends upon the excess temperature. Thus,

$$Q = K(T - T_r)$$

where K is a constant, the "heat exchange coefficient," which is related to the area exposed to dissipate heat, the nature of the gas (air in this case), the thermal contact of the wire with the materials around it, and other factors none too easy to evaluate. In effect, this Q value should be subtracted from the input power P, so that the equation of temperature versus time becomes not a straight line, as above, but rather an exponential curve going from room temperature asymptotically toward the temperature at which the wire heat loss is equal to the heat input.

How Big Is the Bang?

The energy due to compressed gas in a closed vessel is just:

$E = PV \log_e(P/P_a)$

where P_a is atmospheric pressure, P is the pressure of the vessel, and V is the volume. Our balloon contains $^4/_3\pi R^3$ m^3 of gas, where the radius R is 125 mm, so $E \sim 4$ Joules (J), depending upon the value of P chosen. With a balloon, P cannot be much above atmospheric pressure—just a few tens of millibar above. Suppose that $P = \Delta P + P_a$, where ΔP is small:

$E = P_a V \log_e(1 + \Delta P/P_a).$

Since $\log_e(1 + x) \sim x$ for small values of x, the energy available is:

$E \sim V\Delta P.$

With a typical value of 50 millibar, the energy available is around $4/20 = 0.2$ J.

The energy available to make the sound is identical for both an air-filled and a helium-filled balloon. However, the greater speed of sound in the case of the helium balloon means that the energy is released more quickly: the sound is at a higher frequency. You should find, if you have a group of people listening, that the youngsters will hear the helium as louder. Why is this?

Speed of sound $\sim \sqrt{(\gamma RT/M)}$

where γ is the ratio of the specific heats C_p/C_v for a gas, R is the gas constant, T is the absolute temperature, and M is the relative molecular mass of the gas molecules or atoms. Air is 78 percent nitrogen ($M \sim 28$) and 21 percent oxygen ($M \sim 32$), yielding an average $M \sim 29$, while helium has $M = 4$.

So if $v_{air} = 340$ ms^{-1} ($M \sim 29$)

then $v_{helium} \sim \sqrt{(29/4)} \cdot 340 = 915$ ms^{-1}.

The higher sound speed means that the rise in pressure due to the exploding balloon can be quicker with helium. The 0.2 J of energy in the balloon will be released 2.7 times faster from the bursting of a helium balloon, giving the bang a predominant sound more than an octave higher. The human sense of hearing varies with age, with young people being far more sensitive in the higher audio and lower ultrasonic regions from 5 to 15 kHz, which explains why youngsters think that helium balloons make a louder bang.

And Finally . . . The Chain Reaction Balloon Explosions

The balloon detonator goes off only when the switch is pressed. However, there is a simple way in which we could trigger the explosion of one balloon using the explosion of another. A balloon can be equipped with a pair of contacts held apart from touching by the stretched rubber of an inflated balloon. On bursting that triggering balloon, either with a balloon detonator or with a pin, the contacts will close and, if wired to other balloons equipped with balloon detonators, will cause those balloons to explode.

The diagram shows how two triplet "chocolate block" connectors can be connected with steel pins, pulled together with an elastic band. They should be

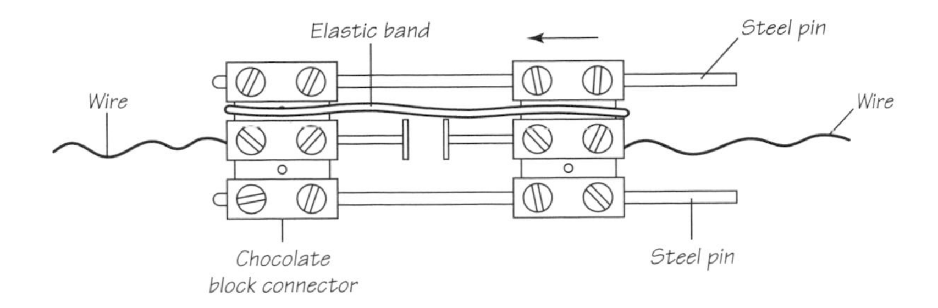

held apart by being fixed to the balloon with sticky tape. The middle connectors are the active ones, which will connect when the chocolate blocks meet. Wires from these can be led off to another battery and one or more balloon detonators. Be careful about sharp objects sticking out of the connectors—otherwise premature balloon explosions will result.

Try setting up a chain of two or three balloons, and then pop the first one. The delays between each balloon exploding add an element of unpredictability to the process. Maybe you can set off a whole cascade of exploding balloons. Would the cascade follow a Fibonacci sequence if each balloon exploded two further balloons?

1 balloon explodes	1 has gone
2 are exploded as a result	3 have gone
4 are exploded as a result	7 have gone
8 are exploded as a result	15 have gone
16 are exploded as a result	31 have gone
32 are exploded as a result	63 have gone
. . . and so on	

What about the thermal delays between explosions while the electric current causes the wire to heat to the balloon's exploding temperature? Are these delays a precisely defined quantity as the simple calculation above suggests? And what about the problem of the power supply: unless you have a heavy-duty source of current, like a lead-acid battery or a powerful car-battery charger, the voltage at the detonators will fall significantly as increasing numbers of wires are switched on. Will the use of extremely thin (high-resistance) wires help? What effect will this have on the buildup of the chain reaction?

REFERENCE

Supplier for balloon exploders: Conwin Inc., 4510 Sperry Street, Los Angeles, CA 90039; www.conwinonline.com.

3 Slowly Exploding Bolts

This is Apollo/Saturn Launch Control, T minus 55 seconds and counting.

Neil Armstrong just reported back.

All the second-stage tanks now pressurized.

35 seconds and counting. We are still go with Apollo 11.

30 seconds and counting. Astronauts reported, feels good.

T minus 15 seconds, guidance is internal, 12, 11, 10, 9, ignition sequence starts, 6, 5, 4, 3, 2, 1, zero, all engines running.

LIFTOFF. We have a liftoff, 32 minutes past the hour. Liftoff on Apollo 11. . . .

CAPCOM: Apollo 11, this is Houston. You're go for separation. Our systems recommendation is arm both pyro buses. Over.

SC: Okay. Pyro B coming armed. My intent is to use bottle primary 1, as per the checklist; therefore I just turned A on.

SC: Separation complete.

CAPCOM: Roger.

PAO: We confirm the separation here on the ground. . . . This is Apollo Control. Apollo 11's velocity now 21,096 feet per second, distance from Earth 6,649 nautical miles.

—Radio commentary on Apollo 11 launch, 1969

When I first heard of exploding bolts, I thought someone was making a joke. They sounded like something from the Acme Products Corp.'s mail order division—the sort of product that Wile E. Coyote would employ in his endless quest to catch the Road Runner. But look back at old broadcasts from the time of the Apollo landings on the moon, and you will find numerous references to them. They were used, for example, for separating the various different rocket stages as they burned out during the launches.

Exploding bolts—now often called pyrotechnics, or just "pyros"—are still used in considerable numbers in space vehicles. The gigantic Apollo moon rockets had hundreds of them. After the Apollo program, NASA made a policy decision to build their new manned space vehicles for the Shuttle program without exploding bolts. So guess how many are used on the Space Shuttle?*

Exploding bolts required, at the time, a large NASA research and development program to make them safe, reliable, and effective, and so a homemade version of a true exploding bolt is not a recommended project. There are ways, however, of achieving a similar effect, using the same Nichrome wire and high-current NiCd batteries we assembled for the exploding balloons.

What You Need

- ❑ Nickel-chromium (NiCr) wire
- ❑ Nickel-cadmium (NiCd) batteries
- ❑ Balsa or wood to join together
- ❑ Glue
- ❑ Wood blocks to construct model building

For the hot cheese wire

- ❑ Spring
- ❑ Hot-melt glue (glue gun)

What You Do

The diagram shows what you need to do. Set a zigzag of NiCr heater wire into the hot-melt glue in the middle of the joint between two pieces of wood that form the "bolt." When the wire is heated by a current, the hot-melt adhesive melts and the bolt pops apart. To demonstrate the system working, you can simply suspend a weight—a kilo or two would be suitable for a piece of 1-cm (3/8-inch) wood—from the exploding bolt.

*The answer: 267! To be fair, very few of these are actually deployed on each launch—most of them are buried inside systems intended only for use in emergencies.

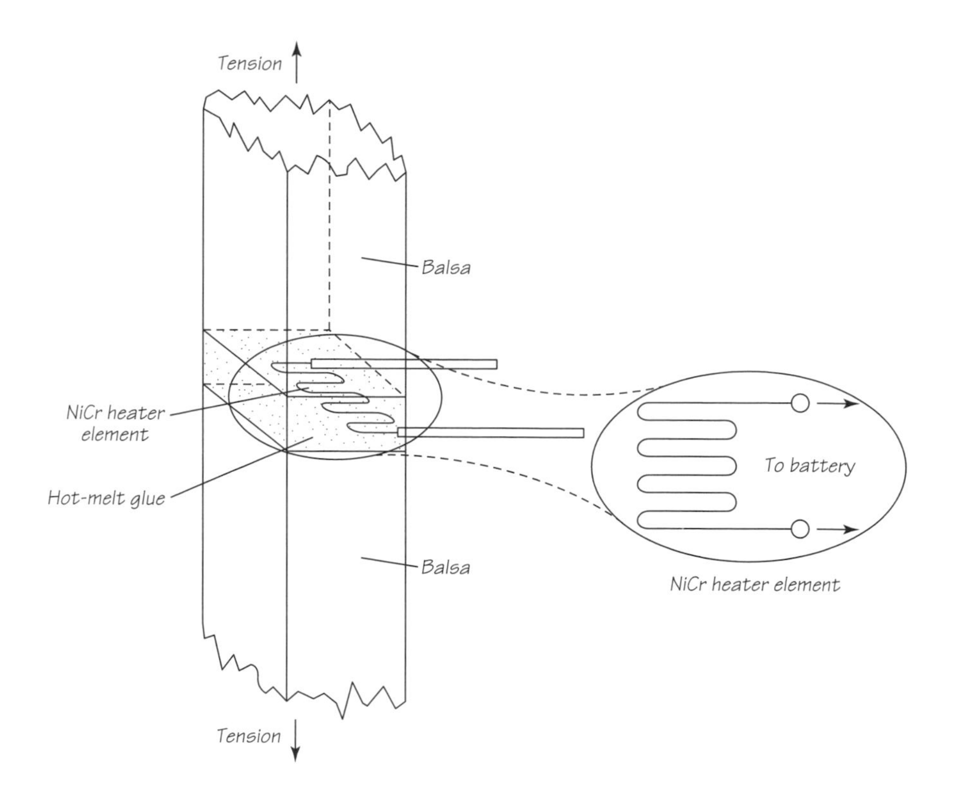

To demonstrate the exploding bolt as the trigger in a system, you could build a two-stage rocket and then install the system between the stages—but maybe this would be allowing the tail to wag the dog. An easier and still satisfying demonstration is to assemble a simple structure of stacked bricks and then knock it down using an exploding bolt. Pin together a parallelogram frame, using the exploding "bolt" as a diagonal strut. Make a matching, noncollapsing thin block—about 25 mm (1 inch) thick for stability—for the other side of the base, then carefully build a towering model building from large wood blocks. You will probably find it necessary to add thin pieces of wood or paper (shims) to one side or the other to keep the building vertical. I found I could stack a few kilos of 25- and 50-mm (1- and 2-inch) blocks up to 1 or 1.5 m high (3 or 4 feet) on top of a 10-mm ($^3/_8$-inch) wood frame, a size of structure that tumbles down with a satisfying thump when you blow the exploding bolt (see diagram).

I have sometimes gathered together hundreds of wooden blocks and asked a bunch of kids to build a tower with them of the maximum possible height. Using blocks no thicker than 50 mm (2 inches), we have built towers—without glue or

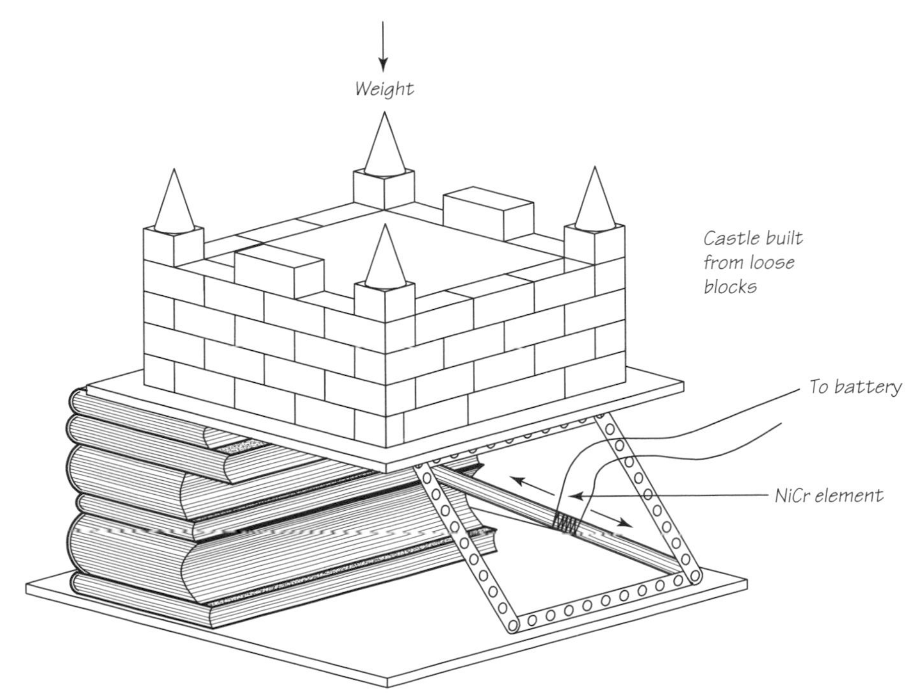

Demo rig keeping NiCr element under tension

other artificial aids—up to about 3.5 m (12 feet) tall. One good strategy is to use a three-column structure, with platforms or long blocks connecting the three columns about every 300 mm (1 foot). Keep the tower platform stages strictly horizontal using a spirit level and thin pieces of rigid material—"shims"—to maintain that horizontality. You will need to use a stepladder to complete the higher parts of such a tower. Beware of standing too close when it collapses—the total weight of the blocks will be considerable.

The Hot Cheese Wire Bolt

The buried wire technique works fine for quite large and strong structures, which is good. But it is not much like the original NASA exploding bolt. Spacecraft exploding bolts are normally under tension, whereas a simple butt joint of glued wood pieces is not, of course, strong in tension.

By contrast, the hot cheese wire relies on a Nichrome wire poised and tensioned to cut through a cylinder of hot-melt glue. The cylinder of glue is considerably stronger than the butt joint because it has no joint faces in it. The cylinder

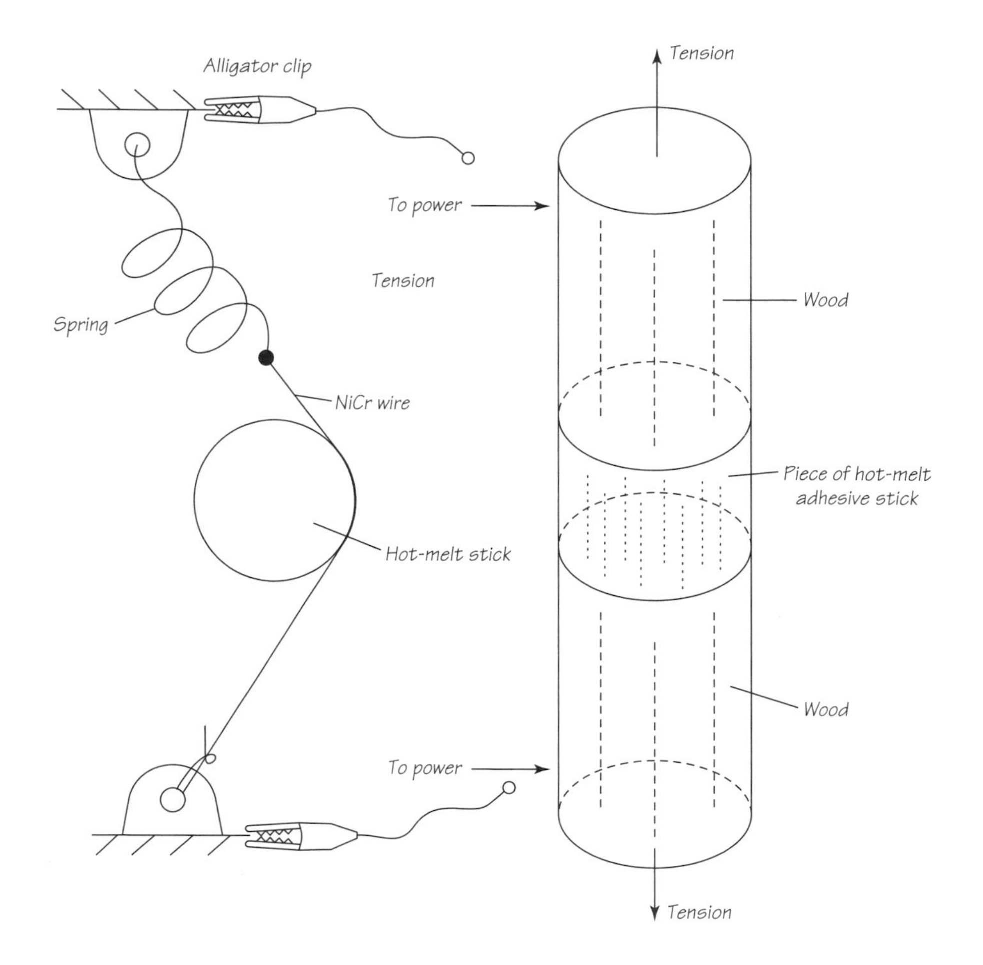

must be installed in a way that ensures it is under tension, as in a suspension bridge cable or in the longer diagonal of a parallelogram under compression. On applying the command current, the NiCr heats, and, as fast as the current can heat up the plastic material to the point where it is soft enough to be cut by the wire, it cuts through the cylinder; the cylinder being in tension, it pulls apart after a few seconds. The diagram illustrates how the system is intended to work. You should minimize the amount of NiCr wire that extends beyond the cylinder of glue, since this length of wire will not do any work of melting and will reduce the current in the circuit.

The calculation of the joint-break speed for the buried hot wire, at least approximately, resembles that for the balloon detonator:

$$E = M_w C_w (T_{soft} - T_r) + M_g C_g (T_{soft} - T_r)$$

gives the energy E needed to heat the buried hot wire, mass M_w, specific heat C_w and the glue in the joint (mass M_g), and specific heat C_g; T_{soft} and T_r are the softening point of the glue and the room temperature, respectively. The power P applied to the wire is given by the electrical power formula $P = VI$, where V is the voltage applied and I is the current flowing. Using Ohm's law, $V = IR$, we have once again $P = V^2/R$, where R is the resistance of the wire.

The time t_{ex} taken to explode is given by the buried wire exploding bolt equation:

$$t_{ex} = R(T_{soft} - T_r)(M_w C_w + M_g C_g)/V^2.$$

Thermal conductivity is not taken into account in this formula. The thermal conductivity of the glue, though, is probably of little consequence. Not so, however, for the effects of the material that the glue joins. With a wood bolt, heat is almost 100 percent confined to the joint glue, and the above equation applies. With an aluminum exploding bolt, however, a much larger power would be needed to raise the temperature enough to cause failure, because the metal will conduct much of the heat applied away from the joint.

The calculation of heat lost through the substrate could follow the analysis of the classic Searle's bar experiment, were it not for the fact that Searle's bar requires that equilibrium temperatures be reached by the apparatus. More complex thermal-wave calculations must be made for the case where the steady state is not reached. The vital parameter here is the thermal diffusivity D_t, which is a function of both thermal conductivity K and specific heat capacity C:

$$D_t = K/C.$$

The appropriate equation is the diffusion equation:

$$D_t(\partial^2 T/\partial x^2) = \partial T/\partial t,$$

which leads to solutions like $T - T_r = \exp(ax^2/t)$, showing heat spreading out a distance proportional to the square root of time, as shown in the graph.

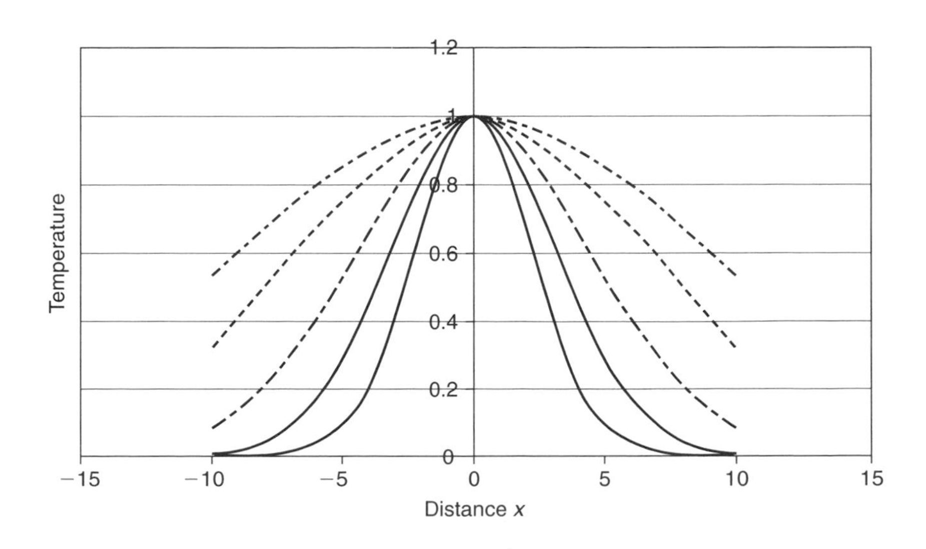

The hot cheese wire presents some different features for analysis relative to both the buried wire and the balloon detonator. The short piece of hot cheese wire does not melt all the plastic at once. It only has to melt the plastic immediately in front of the wire, and it does need to melt all the plastic that lies in the wire's plane. We can neglect, then, the time taken for the wire to heat, but the time taken to heat the glue is given by a similar formula, provided that we take for the mass of glue M_g the volume of glue about a wire's diameter (D_w) in thickness and the full cross-sectional area of the cylinder's diameter D being cut:

$$M_g = \rho D_w \pi D^2/4.$$

We thus can derive the hot cheese wire equation, giving the time t taken for the cheese wire to cut right through:

$$t = R\,(T_{ex} - T_r)(\rho D_w \pi D^2 C_g)/4V^2,$$

which shows a typical $1/V^2$ dependence that we expect and also shows the importance of using a small-diameter wire.

And Finally . . . Gunpowder Glue

You don't have to use hot-melt glue as the thermoplastic material. Many other materials that melt or lose a large proportion of their strength at a defined and reasonably low temperature will serve, particularly those that have a reasonably low thermal conductivity. As discussed above, the thermal conductivity of the substrate around the joint in the buried hot wire is important, while the thermal conductivity of the plastic itself is the vital parameter in the hot cheese wire. Other meltable adhesives are possible. Polymer fibers such as Kevlar are among the strongest known materials in tension, and they are also fairly easily melted. Polypropylene, used in string and ropes, melts even more easily but is still surprisingly strong. What is the strongest material you can find that will function as a slowly exploding bolt? And can you assemble a demonstration based upon that material?

Another material to try might be the adhesive known as "balsa cement." Some of these cements are a preparation of cellulose nitrate polymer dissolved in a small amount of solvent; the polymer sets as the solvent evaporates. Balsa cement is flammable before it sets because of the solvent content. Surprisingly, it remains fairly flammable even when set because the nitrate groups release oxygen as they decompose, and cellulose nitrate will burn slowly, like a rather inefficient firework, even in the absence of air. Old movies were recorded on photographic film made with a cellulose nitrate polymer base that could catch fire easily and was difficult to extinguish once ignited. Old nitrate-stock movies are today kept in metal cans, and often refrigerated as well, to prevent degradation and to minimize the possibility of ignition.

This raises the possibility of using red-hot NiCr wire not just to melt, but actually to ignite the balsa cement. Provided that the film of adhesive is fairly thick—at least 2 mm—it will continue to burn. A considerably larger area of adhesive than can be directly heated by the NiCr wire can thereby be destroyed, allowing a much larger joint to be "exploded." The effect could possibly be enhanced by adding powdered oxidants such as nitrates or perchlorates to the glue—but this would then be a true kind of gunpowder and might present some of the well-known hazards of ordinary gunpowder.

REFERENCES

Exploding Bolts

Bement, Laurence J., and Morry L. Schimmel. *A Manual for Pyrotechnic Design, Development and Qualification.* NASA Technical Memorandum 110172. Hampton, Va.: National Aeronautics and Space Administration, Langley Research Center, 1995.

Nonexploding Bolts

Professional nonexploding bolts are now being manufactured and used in the space industry instead of "pyros." Take a look at Frangibolts, for example, which employ a cylinder of shape-changing titanium-nickel alloy. When heated, the cylinder expands by several percent, enough to shatter a bolt that is compressing it and holding parts together. Further information regarding Frangibolts, made by TiNi Aerospace of San Leandro, California, is available at www.tiniaerospace.com.

New Motors

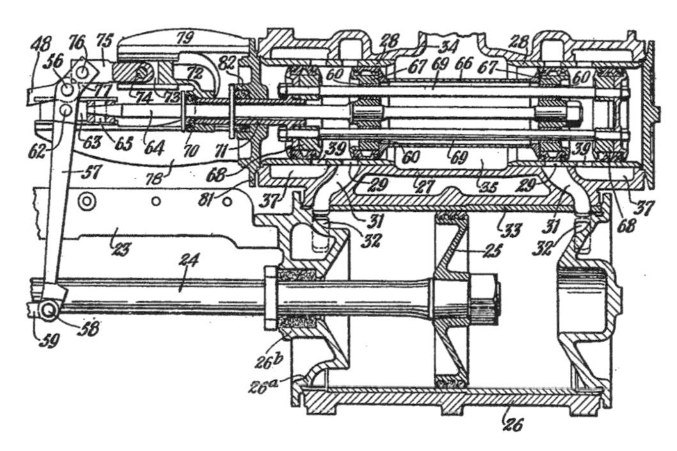

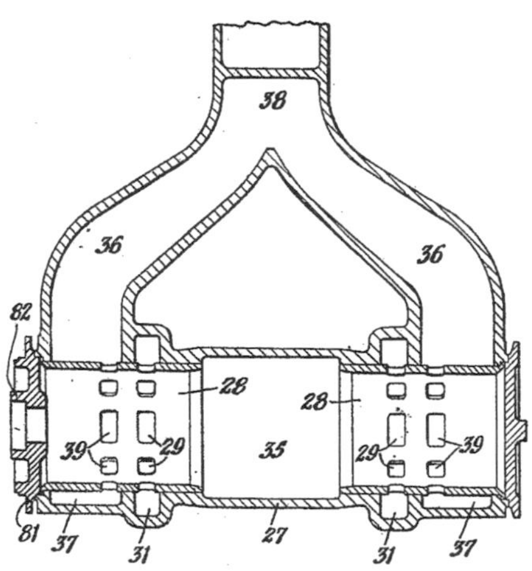

The development of the steam engine was an empirical process . . . this whole process was essentially "hands-on," a technique often referred to in industry generally as "sitting with Nellie."

—P. W. B. Semmens and A. J. Goldfinch, *How Steam Locomotives Really Work*

An engine—or, more accurately, perhaps, a motor—can be described as a device for converting energy, typically heat energy or electric energy, into motion (mechanical work). Our first curiosity from the world of engines is a kind of steam engine. But it is a steam engine that does not employ the high pressure of steam or, indeed, a high pressure of anything. Instead it uses no pressure at all, but rather the reverse of pressure: a vacuum. Its power is generated by expanding the air around us into a vacuum. Although rarely used today, this was a principle widely exploited in the earliest days of the Industrial Revolution.

Almost all electric motors are purely rotary in their action. Here, however, we explore an electric motor that does not rotate but rather goes back to the reciprocating motion of Victorian engines. Reciprocating motors that use a ratcheting action have been making a bit of a comeback in microelectromechanical systems (MEMS), where, at least in subminiature sizes, they may eventually become more widely used than purely rotary motors.

Last we consider a kind of electric motor—a rotary type this time—that has been growing in importance over the last two decades as computer control has become ubiquitous: the stepper motor. Normally a stepper motor is a fairly complex piece of electromechanical design, requiring an even more complex set of driving circuits. But here we strip it down to its bare essentials, with the happy benefit of low cost—to the extent that it must be the world's cheapest stepper motor.

4 Vacuum Engine, or Negative Steam Engine

I sell what all the world desires: power

—Matthew Boulton, pioneer steam engine manufacturer, partner with James Watt, Birmingham, 1780

Power is transmitted in everyday life most often by electricity. There are other means of power transmission such as high-pressure air, high-pressure hydraulic oil, and, on industrial sites, steam. However, electricity dominates: it is the most versatile form of energy. It can be converted efficiently to any other form of energy, something that is not true of other types.

Most kinds of power transmission have a certain degree of tangibility. The rotating propeller shaft of a large truck leaves no doubt that considerable power flows from the engine at the front to the axles at the rear. Wander near a high-power electric system, and you'll readily hear the low but insistent hum at the 50 or 60 Hz line frequency; sometimes you'll even feel the hairs on your body react. The apparently inert wires and cables of high-power electric systems can produce huge and mortally dangerous flashes and sparks if they are disturbed. Similarly noisy and spectacular gas jets signal the presence of even small leaks in compressed air or steam systems.

By comparison, the transmission of power through a vacuum in a pipe seems a peculiarly intangible concept. How can power be apparently transmitted by nothing? But in this project we show that a vacuum can indeed transmit power, and that we can demonstrate a motor rather like an old-fashioned steam engine, an engine that can turn the power transmitted by a vacuum in a pipe into mechanical energy.

The Industrial Revolution that transformed the Western world, starting about 1700, needed mechanical power. At first, increased use and more efficient designs of watermills and windmills could provide that power. But it gradually became evident that the continuous power which steam could provide was going to be needed. It is easy to appreciate the expansive force of steam when you see a kettle boil. However, none of the early steam engines used that expansive power. Instead they used atmospheric pressure (they became known later as "atmospheric" engines), with the steam being used to create a vacuum so that the atmosphere could push a piston. We might today, perhaps less accurately, call them vacuum engines.

There have been times when vacuum power transmission has been used. Perhaps the first example was the system used by Matthew Boulton and his partner James Watt. Near the Boulton and Watt engine factory in Birmingham, England—the world's first engine factory—was the Boulton and Watt mint, a coin factory operated by one of the company's own engines. Engineer John Southern devised a system in which a steam-driven vacuum pump partially evacuated a huge pipe, known then as the "spirit pipe." Individual coin presses were powered by cylinders and pistons connected to the spirit pipe.

Since the time of Watt and Southern, vacuum power distribution has occasionally resurfaced in different places. Vintage automobiles from the 1920s on were sometimes fitted with a kind of vacuum engine to operate the windshield wipers, using the vacuum from the gasoline engine's inlet manifold. It cannot have been an ideal system: if an engine turns over slowly, the vacuum from the engine would decrease and the wipers would operate more slowly. If you were driving one of these old cars and saw an approaching hazard, you would naturally slow down. And just when you needed more wipes of the windshield to see what was going on, the opposite would happen: the wipers would slow down and you would be left peering through rain-swept glass at exactly the wrong moment!

Today this principle is still being used in at least one application (albeit rarely): vacuum cleaners. In some models of cylinder vacuum cleaners with a rotating brush, the brush is powered by a simple turbine device that is turned by air sucked into a vacuum created by a centrifugal fan in the cylinder.

Our vacuum engine is a "steam engine" type of device. Unlike most steam engines, however, it does not require a fully equipped workshop with lathe, milling machine, and so on. Neither does it need the thousandth-of-an-inch accuracy required of a working model steam engine. The vacuum engine only requires a few hand tools, pieces of wood, plastic tubing, and easily obtained metal hardware, and you don't need to make anything more accurately than within a millimeter. You won't burn your fingers, either—because you don't need steam! It is also easy to make—you can probably assemble one in an afternoon. Nevertheless, it well illustrates all the main working principles of steam engines: piston and cylinder, crank, flywheel, valve gear, and valve timing. Take a look at books like that of Semmens and Goldfinch if you want to know more about steam engines.

What You Need

- ❑ Vacuum cleaner (ideally the horizontal cylinder kind)
- ❑ Short section of 18-mm (3/4-inch) hose
- ❑ 300-mm-long, 32-mm-diameter plastic pipe
- ❑ ca. 150-mm-long, 31-mm-diameter round section of wood to fit snugly in pipe
- ❑ Flywheel pulley from an old washing machine
- ❑ Brass rod that will roughly fit the hole in the flywheel
- ❑ Metal shaft and brackets
- ❑ Conrods (e.g., 8-inch by 1/2-inch Erector set strips)
- ❑ Wood pieces
- ❑ Electric drill
- ❑ Bolts and nuts
- ❑ Hot-melt glue

How to Build a Vacuum Engine

The basic idea of the vacuum engine is that a piston is propelled up and down to push a crank that connects to a flywheel. The piston is activated by atmospheric pressure on its connecting rod (conrod) side, with periodic pulses of vacuum applied to its piston-head side. The pulses of vacuum pressure are applied by intermittently connecting the low pressure from a vacuum cleaner to the piston. The intermittent connection is made by a slide valve. The valve is synchronized to the flywheel rotation and hence to the piston movement, by being actuated 90 degrees out of phase with the piston in terms of flywheel position.

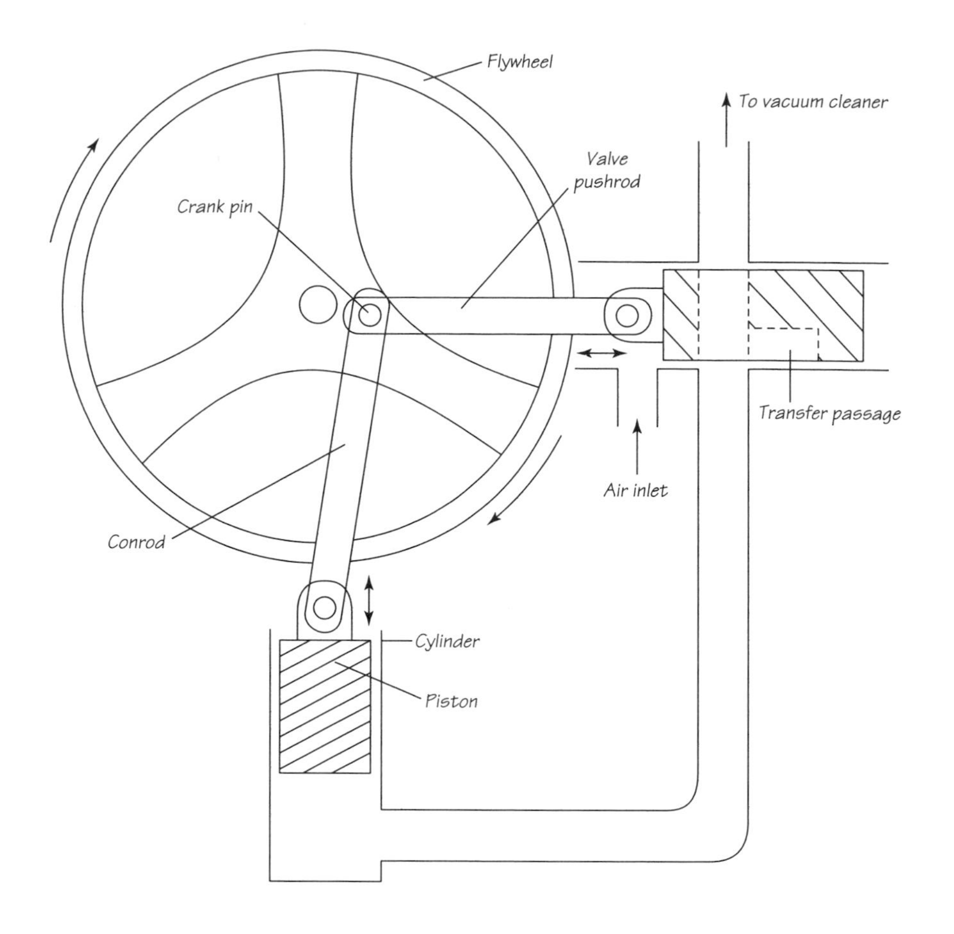

I found a piece of wooden dowel that fit snugly inside the drainpipe I had chosen. I then used this rod and pipe for both the piston and the slide valve. I suggest that you aim for a piston that is about 1 mm smaller in diameter than the cylinder, both for the piston and for the slide-valve assembly. Try to find plastic pipe that is close to precisely round. (You will find occasional pipes or sometimes even entire batches that are appreciably noncircular; perhaps they have been squashed in storage or loading at the factory or supplier.)

The piston, if it is the right size, needs no preparation at all other than to bevel the edges and to screw on the conrod bracket. The slide valve and its cylinder are more complicated. The cylinder needs two or three holes (an air inlet hole is optional) as shown in the diagram, which all need their edges smoothed. You must drill through the valve body for the vacuum port and then make a slot with a chisel for the transfer port. The transfer port allows air into the drive cylinder after it has completed its power stroke.

I have two suggestions for alternative, simpler slide-valve designs: First, you can omit the air-inlet hole and the transfer port channel, relying on air leaking around the piston and valve. Second, you can omit the air hole and transfer port from the valve body and also cap its end. You can now switch the valve on and off with a simple cylindrical piston (exactly like the power piston), by arranging that the piston just uncovers the holes in the valve body as it reaches top dead center (TDC). You will need to cap the end of the valve cylinder in this design too.

I used Erector set parts to construct the crank plate and light steel strips for the conrods. The pipe work was completed with a washing-machine drain hose, which is typically a fairly generous-bore 18-mm (3/4-inch) corrugated pipe. You can minimize the Erector set parts by making your own bearing for the flywheel and fitting the crank pin directly into a small hole drilled into the flywheel. The bearing for the flywheel can be made using a piece of 6-mm (1/4-inch) steel and a piece of brass rod around 15 mm (5/8 inch) in diameter, glued with epoxy adhesive into the center of the flywheel central hole. Bore out the middle of the brass rod with a 6-mm (1/4-inch) drill, deburr it if necessary with an oversize drill bit or just a sharp knife, then run the drill up and down it a few times until the rod will fit snugly but freely rotate around the 6-mm (1/4-inch) steel rod.

The position of the cylinder on the base plate is not critical. The position of the valve body, however, is more sensitive: it must just begin to open to vacuum when the piston is closest to the flywheel (the position conventionally known as TDC).

You must ensure that every part can move freely. Check that the edges of the holes in the valve piston cylinder are smooth and that the pivots on the pistons and the crank are not binding. If rotated vigorously by hand without the vacuum applied, the engine should turn over at least three or four times. If you find that the engine slows more quickly than this, you should check for excess friction in one of the parts.

What You Do

Without a plentiful supply of vacuum, your engine won't work, so make sure that your vacuum cleaner has powerful suction. The stronger the vacuum—meaning the larger the negative pressure relative to atmosphere—the better the vacuum engine will perform. If you hold any doubts concerning the performance of the vacuum cleaner, try to find some means of measuring the negative pressure it produces. The flow rate that the vacuum cleaner can produce is rather less important, as the flow rate needed by the vacuum engine is fairly low and, unless your fabrication of the device is more precise than I have suggested, much of the air flow will go to supplying leaks rather than to propelling the engine. If your vacuum cleaner has a low flow rate, you can still operate a vacuum engine, but you must make the piston and valve pieces a tighter fit within their cylinders.

Now position the flywheel just a little past the TDC. Apply the vacuum. With luck, you should find that the flywheel should begin to turn of its own accord, rushing down toward bottom dead center (BDC) and then beginning to slow down. But it should be going just fast enough to rotate one complete revolution at low speed, after which the process can repeat. The next time the engine will reach TDC a little faster, and the flywheel will complete its revolution more quickly. With the dimensions given here and a reasonably powerful vacuum cleaner, your vacuum engine should build up in speed until it is whirling around at 300 to 400 rpm or more.

How It Works

The vacuum engine works by atmospheric air pressure. When the flywheel is at TDC, air pressure is the same on both sides of the piston, so no force is applied. With the flywheel turned a little, so that the valve opens to the vacuum, air is removed from underneath the piston. With no air pressure below but atmospheric air pressure above, the piston is forced downward.

Curiously, in the engines I have tried, the rather rough-and-ready fit of the wooden piston to cylinder may help, in that the air inlet and the transfer passage in the valve gear did not seem to be necessary. As mentioned earlier, this means that you can simplify the engine and use a piston as the slide valve. With a better standard of construction, you will need a proper slide valve with an air inlet.

Troubleshooting

Like the original steam engines, your vacuum engine may need a little adjustment before it will run properly (or perhaps run at all). You do need to ensure that all parts run smoothly and are lubricated with a little light oil such as bicycle oil.

The highest friction forces, assuming that all your components are smooth running under freewheel conditions, will be developed when the vacuum is applied to the slide valve. With a fairly close-fitting slide valve, this force will be reasonably low. If, however, like me, you started with a rather loose-fitting slide valve, you will find that it tends to bind. What is happening here is that the valve piston is being pulled hard against the valve cylinder because of air pressure on the side opposite the vacuum cleaner connection. Some oil may fix the problem. If the fit is really loose, worse than 1.5 mm smaller than the cylinder, then it may be necessary to start again and make another slide-valve piston with a better fit. A simpler solution if the fit is not too bad is to glue a cap onto the end of the slide-valve cylinder. This blocks half the flow of leakage air to the slide valve and reduces the force needed to operate the valve. (Thanks to the kids at the Saturday Activity Center in Guildford, U.K., for that tip.) Of course, if you have used the simplified piston-style slide valve, then you will have blocked off the end anyway.

THE SCIENCE AND THE MATH

The easiest way to understand the engine's cycle mathematically is by plotting a sine wave to represent the piston and the valve-gear positions, which is what is shown in the graph. (The sine curve is an approximation, but if the conrods are long it is a good approximation.) Once the engine has turned about 45 degrees past TDC, the vacuum is applied and the engine will be forced around by atmospheric pressure until it reaches about 140 degrees past TDC, when the vacuum will be cut off once again and leaks will refill the piston/cylinder chamber, allowing the engine to run relatively freely past BDC at 180 degrees past TDC, and then freely to TDC once again.

Note that power is only being applied to the piston for 90 degrees of the 180 degrees past TDC for which power could be usefully applied. By readjusting the length of a valve rod or the width of the valve slot, you could extend the period over which power is applied. You could even extend the valve opening to a bit before TDC, because once the engine is running fast it will not be stopped by a momentary reverse force in each cycle, and this will ensure that the slide valve is more fully opened during the power stroke.

We have thus proved that "nothing" can transmit power. But what pressure of nothing is best? What is the optimum pressure for transmission of power? A pipe will not transmit much power if the pressure inside is very close to atmospheric pressure: when the pressure is equal to the atmospheric, there is no differential pressure drive, and no power would be transmitted at all. So the optimum pressure must be lower. At the other end of the scale, the power that can be transmitted at very low pressures must be low too: at an extremely low pressure, the mass flow through the pipe will be greatly restricted. In an ultra-high vacuum, almost zero mass of air would be moved along the pipe, and almost no power could thus be transmitted. The maximum amount of energy that a vacuum pipe can transmit occurs, therefore, when the pipe is at intermediate pressures.

The energy E_{comp} stored in a compressed gas that is expanded from pressure P_i and volume V_i to

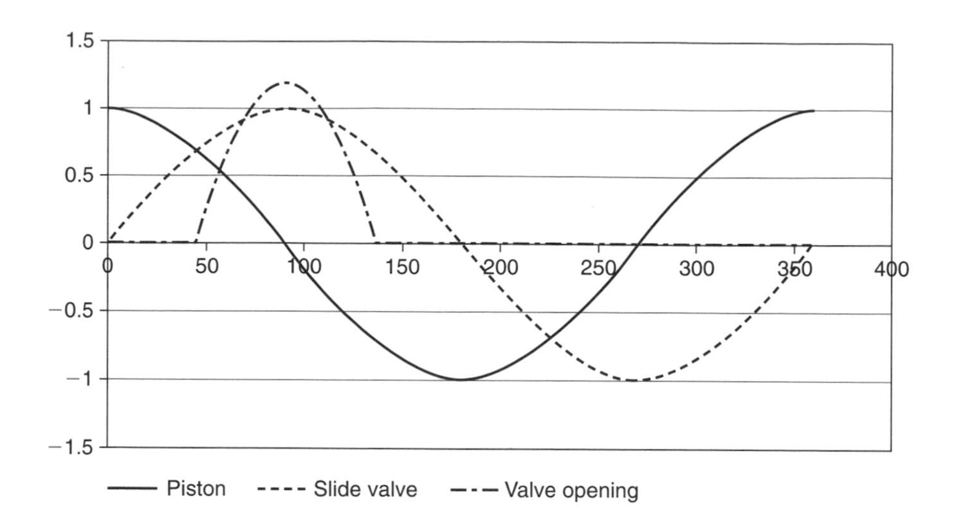

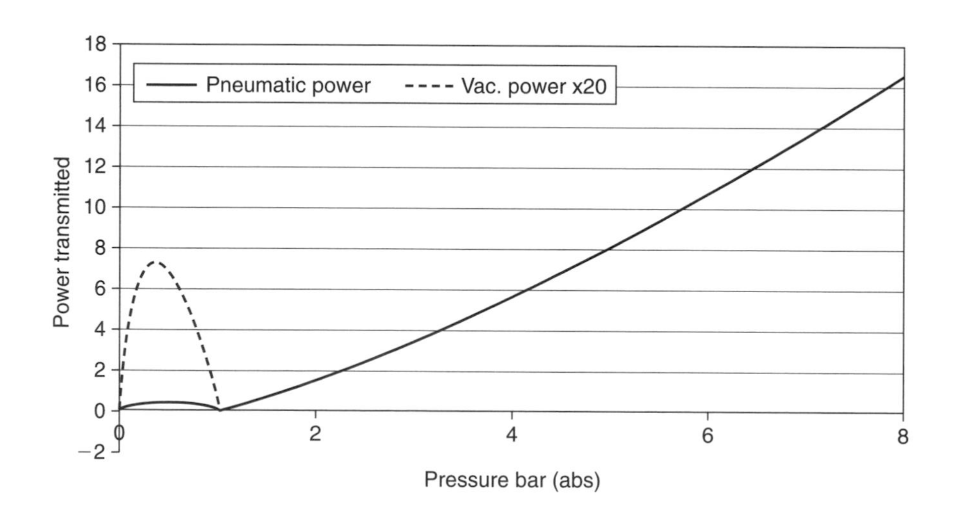

final pressure P_f and final volume V_f (see this book's first project, the exploding disk cannon) is given by:

$$E_{comp} = P_i V_i \log_e(P_i/P_f).$$

If we assume that the pipe is at absolute pressure P, and that the flow rate is Q, then the power transmitted in a vacuum pipe is governed by what we might call the vacuum engine transmission equation:

$$P_w = PQ \log_e(P_a/P)$$

where P is the pressure of supply and P_a is the atmospheric pressure. On the basis of this equation, the power transmitted, P_w, increases linearly with the volume flow rate Q. However, the pressure dependence is curious: as P gets closer to P_a, the power transmitted will tend to zero with the logarithmic term, while as the pressure P tends to zero, the power also tends to zero, but with the linear term.

The plot in the graph contrasting pneumatic and vacuum power transmission shows that the optimum pressure from a vacuum source is about $P = 0.4P_a$. This is a surprising result: there is a value of vacuum pressure that is naturally favored for power transmission. Other power transmission systems have no such favored parameters. The voltages used in electrical transmission, for example, should be as high as possible, consistent with safety in the case of domestic supply and in commercial transmission by the quality of the insulation on the wires. Similarly, the air pressures used in pneumatic systems are fairly arbitrarily chosen to be 100 psi or 7 barg, at which values components like plastic flexible piping and valves are inexpensive; as far as efficient transmission of power goes, the highest possible pressure is indicated. There are also other considerations, of course, such as flow rates, pipe sizes, pressure drops in pipes, and so forth.

If we compare vacuum power at the optimum 0.4 bar absolute with conventional pneumatic power at 7 bar gauge (8 bar absolute) using the same volume flow rates, we find that the pneumatic system will transmit nearly fifty times as much power. The second graph extends the vacuum-power graph to include both vacuum and pressure power transmission. To reach a reasonably large power transmission with the vacuum system we need to use larger pipes giving larger flow rates. With laminar flow patterns, we might expect flow rates at a very low pressure differential, to depend upon r^4, approximately, since

$$\text{Flow} \sim \text{Area} \cdot v \sim r^2 v$$

where the mean stream velocity v goes as r^2 for laminar flow. However, this formulation is unrealistic for our situation. For well-developed turbulent flow, where $v \sim r$, we will have

$$Q \sim r^3.$$

So, by increasing the tube radius by a factor of about 45 to the power 0.33, or 3.5, we should be able to provide the same power as a pneumatic system. The work done by a pneumatic system with 6-mm (1/4-inch) tubing could be done by a vacuum system with 22-mm (7/8-inch) tubing.

ABSOLUTE, GAUGE, AND DIFFERENTIAL PRESSURE

Absolute pressure is the pressure exerted by gas molecules on a surface, simply proportional to the number of gas molecules per unit volume. With no gas molecules, there is an absolute pressure of zero. This is the pressure measure encountered in most scientific work, used to calculate mass of gas in physics problems and moles of gas in chemistry. The gauge pressure is the standard engineering pressure measured by a simple dial gauge such as the Bourdon gauge. It measures the pressure relative to atmospheric pressure. On a day with an atmospheric pressure of 1,010 millibar, if the absolute pressure of a helium gas supply is 1,200 millibar, the gauge pressure will be 190 millibar. Similarly, differential pressure is the difference between two points of varying absolute pressure. Thus the differential pressure between a helium supply at 1,200 millibar and a nitrogen gas supply at 1,500 millibar will be 300 millibar.

And Finally . . . Spherical Balls and Pelton Wheels

You can experiment with many adjustments that might improve the vacuum engine's efficiency enormously. Sealing the piston and slide valve more effectively to decrease leakage, or increasing the size of the piston and the slide-valve passages, will greatly increase the power generated. An air inlet on a vacuum engine that was well sealed would itself be a source of vacuum and could be used to operate another, lower vacuum engine; the combination would constitute a compound vacuum engine analogous to the compound reciprocating engines of yesteryear, with their multiple stages of steam expansion. The compound engine principle has been carried to even greater heights today in turbines, which can have a dozen or more stages, all crammed together or sharing a coaxial shaft.

You don't need to be restricted to a cylindrical piston and cylinder. What about using a spherical ball as the piston, for example? A spherical-ball piston eliminates the need for a "little end" bearing between the connecting rod and the piston, since the ball can be rigidly mounted on the top of the conrod. Or what about a Pelton wheel, as used in vacuum cleaner engines? You could extract the turbine wheel from a discarded hose system and connect it with a reduction transmission system to convert its high-speed rotation to a more manageable speed.

How much power can you get from the vacuum engine? Could a "vacuum manifold" be used for powering engines in several different places? Or would the pressure transients due to one engine interfere with the running of another? Would similar engines on the same manifold tend to synchronize, rather like pendulum clocks on the same shelf?

Vacuum power is clean: it sucks away any contamination as it provides power. Could it be a useful power transmission system in the future, in places like the wafer fabs of the electronics industry or operating rooms, where ultimate cleanliness is necessary?

REFERENCES

Semmens, P. W. B., and A. J. Goldfinch. *How Steam Locomotives Really Work.* Oxford: Oxford University Press, 2000.

Uglow, Jenny. *The Lunar Men: The Friends Who Made the Future, 1730–1810.* London: Faber, 2002.

5 Rope Ratchet Motor

> Let us try to invent a device which will violate the Second Law of Thermodynamics. Let us say we have a box of gas at a certain temperature, and inside there is an axle with vanes on it. All we have to do is to hook on the other end of the axle a wheel which can turn only one way—the ratchet and pawl. Then the wheel will slowly turn, and perhaps we might even tie a flea onto a string hanging from a drum on the shaft, and lift the flea! . . . [But] the fundamental deep principle on which all of thermodynamics is based . . . tells us that if the temperature is kept the same everywhere our gadget will turn neither to the right nor to the left.
>
> —Richard P. Feynman, *The Feynman Lectures on Physics*

You might think that electric motors must have magnetic parts that go round and round. Not so! Early electric motors, from soon after Faraday's experiments in 1831, included reciprocating types based on solenoid coils and cranks. More recently, Eric Laithwaite, a professor at one of my *almae matres*, Imperial College in London, pioneered linear electric motors. Linear motors assemble coils, iron, and magnets into a kind of "magnetic river" that can both accelerate and levitate a vehicle. Laithwaite godfathered the first NASA Maglifter tests, which might ultimately lead to much more affordable spaceflight. Maglifters, small-scale prototypes of which already exist, would be like gigantic catapults, using magnetic fields instead of elastic to propel projectiles to speeds many times that of sound.

You might also think that a reciprocating motor can only convert its action to a rotary form by means of a crank. Not so! In the early days of steam, James Pickard, a Birmingham, U.K., neighbor of the more famous Boulton and Watt, held a patent on cranks for steam engines. Boulton and Watt and the other manufacturers quickly came up with alternatives like the chain and beam and the sun-and-planet gearwheel system to get around Pickard's patent. Today reciprocating motion is converted to linear motion and vice versa by a plethora of ingenious mechanisms from cams and swash plates to Z-cranks—just take a look in William Steeds's mechanical engineering reference listed at the end of this project.

The humble crank remains ubiquitous in the internal combustion engines of cars. In this project, however, we use yet another device: a ratchet. It is, however, an unusual kind of ratchet. First, it is made not of rigid material but rather of flexible thread. Second, it does not have to move one tooth at a time but is capable of tiny movements per cycle, because it uses frictional forces rather than relying on a toothed ratchet wheel.

What You Need

- ❑ Aquarium air pump (the oscillating-magnet–diaphragm-pump type)
- ❑ Drum on a shaft (e.g., a thread reel)
- ❑ Thread
- ❑ Elastic band

What You Do

The rope ratchet motor works by using a reciprocating electromagnet to pull repeatedly—fifty or sixty times per second—on the end of a string. The string is wrapped around a pulley drum and then fed to an elastic band to give it some tension. Surprisingly, perhaps, the string-and-drum system acts as a pawl-and-ratchet wheel. It allows the electromagnet to pull the drum around when it moves in one direction but also allows the string to slip, to enable the electromagnet to recover its starting position when it moves in the other direction with-

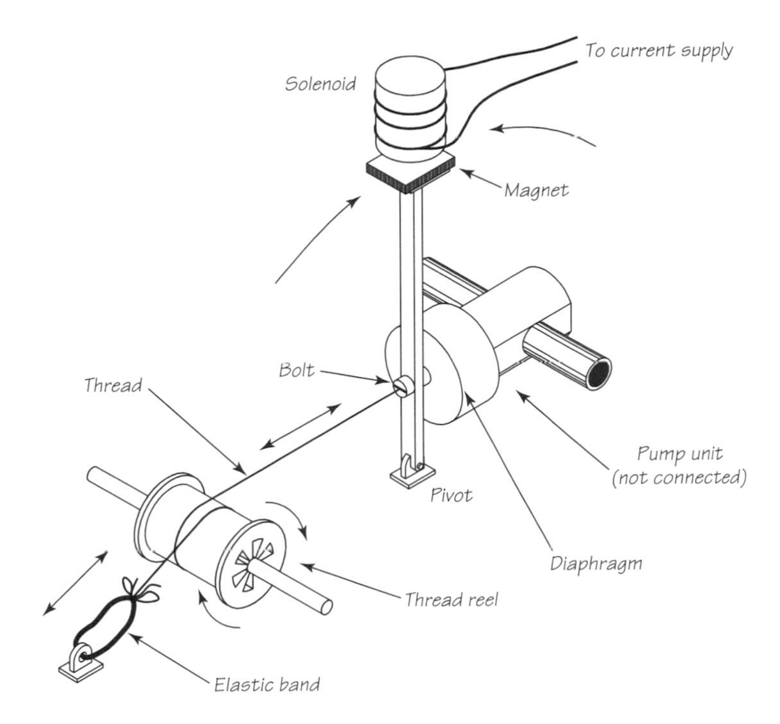

out moving the drum. Thus a to-and-fro action by the electromagnet is converted into a quasi-continuous rotation of the drum.

First, dismantle the aquarium pump so that you can modify it. Because you will not be using the device as a pump, the diaphragms can be removed, although you may find it difficult to keep the magnet stationed accurately above the electromagnet. If this is the case, simply leave the diaphragms in place.

The tension on the drum string must be carefully regulated to optimize the speed and torque of the motor output. You can do this by tensioning the elastic band correctly. If you fit a screw to the elastic band, you can make it continuously adjustable.

Test pull the magnet to one side and then the other, checking that it pulls the drum around in the correct way and allows the string to slip in the other, thus acting as a ratchet. Then power up and see what happens. With luck, the magnet will oscillate in a blur, sixty times per second (or fifty, depending upon your current's frequency), and the drum will rotate perhaps twice per second. If it does not, try changing the tension in the elastic band or repositioning the magnet. The magnet needs to be in the oscillating field of the electromagnet, however, and if you use a comparatively strong elastic band on the string, you may pull it too far away from this central position.

WARNING

Do *not* interfere with the 110- or 220-volt wiring part of the pump. This kind of aquarium pump works by arranging a magnet on a hinged lever above the gap in a horseshoe electromagnet connected to the domestic AC electric power. As the AC reverses polarity, fifty or sixty times per second, the magnet is pulled first one way and then the other way. This action is normally used to push the diaphragm on the pump unit up and down, check valves allowing this up-and-down motion to be converted to one-way air flow.

You may find that the drum rotates only when it has a braking force on it; this is because it is simply oscillating to and fro without the string slipping on the return stroke. This will not matter when the rope ratchet motor is doing some work, but it is annoying not to be able to demonstrate it as freewheeling. Try cleaning off any oil you have put on the axle by using solvent, to give the drum a little more friction, or applying some small braking action like a felt pad. This might be one of the few examples of a motor that works better with sand in its bearings!

If you put a load on the drum by braking it near its axle with your fingers, you should find it surprisingly difficult to stop. By contrast, you can stop the magnet's oscillation with almost no force at all: the rope ratchet works like a downshift transmission. You can pull substantial loads by winding a string around the thread reel axle. To pull the maximum load, you will need to optimize the elastic band tension.

THE SCIENCE AND THE MATH

The theory of the rope ratchet depends upon integrating the tension and friction forces along the string. The force F_p that the ratchet can exert when pulling is given by:

$$F_p = F_o \alpha^L$$

where F_o is the tension from the elastic band, α is a constant depending on the string and drum properties, and L is the length of thread wrapped around the drum. Even for a smallish length L—just a turn or a turn and a half—the value of α is such that F_p will be large: there is such a large force that the string cannot slip, and when the magnet pulls on the string, the drum will be rotated.

By contrast, when the string tension is reversed, the maximum force that can be exerted by the elastic band is only F_o, which will be much smaller than F_p. Even this small force will not be exerted, because although the tension from the magnet side is multiplied by the above factor, the movement of the permanent magnet has now slackened off the string to almost zero tension, and the input tension F_o is very low. Hence, as we've seen, the magnet will pull the drum around when it tensions the thread and leave the drum fixed when it slackens the thread, the inertia of the drum holding it still while the elastic band moves the string back.

The extremely high frequency (50/60 Hz) of the rope ratchet motor's operation means that there are dynamic effects which this static analysis does not explain. In effect, the rope ratchet is sending a wave —like a pucker in a carpet—along the rope, rather than tightening and loosening the entire length of rope all at once. The overall effect, however, is still the same. When a pucker travels down a carpet, the whole carpet moves along by a distance equal to the extra length of carpet in the pucker relative to a flat piece of carpet the same length (see the diagram).

It seems likely to me that the release process of the rope ratchet motor involves a transverse movement of the string more than a longitudinal wave that changes the tension. Look carefully, and I think

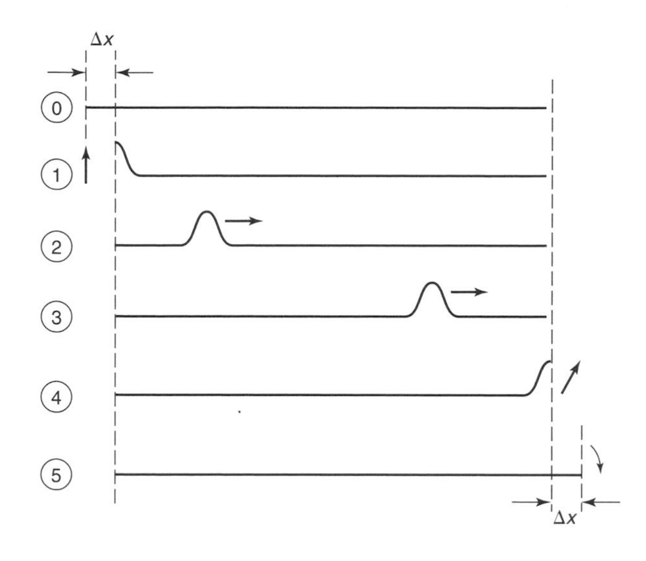

you will see the string flapping outward slightly at each cycle, at least on the beginning of its path around the drum. How fast a transverse wave—or pucker—travels along a string depends upon the tension in the string. The classic equation for the wave velocity v_T is that also used in calculations related to stringed instruments:

$$v_T = \sqrt{(T/\rho)}$$

where T is the tension and ρ the mass per unit length. Stretch the string tighter, and the velocity is higher; use a heavier string, and it is lower. In our case, applying this equation is complex, for the string's tension will vary from place to place on the drum and from moment to moment in the reciprocating cycle of the electromagnet. But we can predict that it will probably be an advantage to use higher values of tension in the higher-speed operations of the rope ratchet motor: the higher tension will, at least within the limits of the square-root law, allow the pucker to travel around and release the string, allowing it to slip faster. Let's put some numbers in: if $T = 1$ newton (N) (an average around the drum), and $\rho = 0.25$ g/m, then we have $v_T = 2$ ms^{-1}. Even if we put a tough elastic band on the device, with $T = 10$ N, then the wave speed will be only 6 ms^{-1}, a rather modest speed. In one cycle of our electromagnet, which takes only 20 milliseconds, a pucker on the connecting string will travel only 40 mm at 1 N, and only 120 mm at 10 N. So the electromagnet that pushes forward a pucker barely has time for the pucker to travel around the drum before the electromagnet reverses its motion and tries to tighten the string on the drum once again.

By contrast, the pulling process of the rope ratchet motor involves more of a longitudinal increase-in-tension wave along the string rather than a transverse wave. A longitudinal wave implies that a different wave-speed equation needs to be considered. Longitudinal waves in a string travel at a speed v_L given by:

$$v_L = \sqrt{(E/\rho)}$$

where E is the Young's modulus for the string and ρ is the density (provided that the string is straight). This formula looks similar to that for the transverse waves, but don't be fooled: it is rather different. The longitudinal wave speed does not depend upon the string tension at all. Provided that the string has sufficient tension on it to keep it straight, the speed of longitudinal waves along it depends only on the string material's modulus and density. Let's run some numbers: suppose that our string is made of nylon, with $E = 2$ gigapascals and $\rho = 1{,}150$ kg/m^3, then we will have a wave speed of around 1,320 ms^{-1}. This is about four times the speed of sound in air and 200 times faster than the calculated speed of the transverse wave on the string in our calculation above. This speed is so fast that we can rely on a longitudinal pull from the electromagnet to appear more or less instantly all around the string wrapped around the pulley drum.

And Finally . . . Faster and Slower Rope Ratchets

It may be possible to raise the frequency at which the rope ratchet motor runs in a simple way. If you replace the driven magnet with a piece of soft iron, it will be pulled toward the solenoid's electromagnet twice as fast—at 100 or 120 Hz, rather than at its normal 50 or 60 Hz frequency. It will be attracted to the electromagnet on each peak, positive or negative, of the current, rather than repelled by a positive and attracted by a negative peak. (A bridge rectifier on the power source, with which the diodes reversed all the negative peaks to positive, might be an alternative way of achieving the same effect. But you need to be experienced with electricity to try this, and the inductance of the electromagnet may result in the electromagnet getting hotter than normal.)

What about a DC-pulse-powered rope ratchet motor? This variation could be programmed to run at different frequencies and with different stroke sizes, to optimize the combination of speed and torque. A simple oscillator like a relay oscillator* could be used to try this out, with different supply voltages to vary the power in each pulse and with different capacitors to program the pulse frequency. This would in effect be a kind of stepper motor, which neatly brings us to the subject of our next project.

*A relay with a normally closed (NC) contact can easily be converted to an oscillator: just wire the power to the relay coil via the NC contact. Whenever the contact opens, current is cut to the coil, which closes the contact again. This oscillation can be slowed down and made more reliable by adding a capacitor of 100 to 1,000 μF across the relay coil. With a double-pole relay, you can use the spare pole to operate the changeover contacts needed to make the stepper motor go.

REFERENCE

Steeds, William. *Mechanism and the Kinematics of Machines.* London: Longmans, Green, 1940.

6 The 50-Cent Stepper Motor

Science owes more to the steam engine than the steam engine owes to science.

—Lawrence J. Henderson, physiologist and science historian, 1917

The stepper motor is the rather complex type of motor at the heart of much modern equipment. Computers, printers, CDs, DVDs, photocopiers, and countless other machines right up to industrial robots all rely upon it. The stepper motor is designed to respond digitally, which of course makes an excellent match with computers. Send a command to move 713 steps to the interface circuit, and that circuit will pulse the several sets of coils on and off 713 times until the motor output shaft has carried out the instruction.

The stepper might superficially resemble an ordinary motor—it is just coils, iron, and magnets, after all. It differs, though, in several ways. First, it is specifically designed to be able hold a stationary position. Most ordinary motors can't remain stationary; this project relies on a particularly simple kind of ordinary motor that does happen to have some position-holding ability. Second, the stepper motor is not, like an ordinary motor, designed to run continuously when you supply current; instead it is intended to move one step and then stop. Last, unlike ordinary motors, which connect directly to an ordinary electricity supply like a battery, stepper motors are useless without their matching driver circuit. (The sidebar for this project describes what that driver circuit does.)

In the early days of power electronics, stepper motors were expensive devices, and very large ones still are. Mass production has brought the price of small steppers down to only $10 or $20, which is why they are so widespread, even in simple computer printers costing only $70. Happily, we can demonstrate all the principles of a stepper motor and all its complex sets of coils and driving circuits with just a simple capacitor and a simple 50-cent motor.

What You Need

- ❑ A small (10 mm × 10 mm × 20 mm) two-magnet, three-coil DC motor, such as the kind used in countless motorized toys
- ❑ Lightweight plastic wheel to fit on end of motor
- ❑ Changeover switch, ideally a microswitch, 5A current rating
- ❑ Capacitor
- ❑ Battery
- ❑ Wires

For the AC Jiggler

- ❑ Transformer (e.g., 12-V AC output)
- ❑ Resistor (correct value to be selected, but probably from 200 to 2000 ohms)

What You Do

Suitable motors should be available from stores that supply small electronic parts, or you can order one from an educational supplier for as little as 30 cents. A simple motor of the type we need has two magnets and three iron-cored coils on its armature or rotor (the rotating part). You should find that when you rotate it, it tends to stop in one of six positions. It should just click into place, almost as if it had six mechanical detent stops. If you fit a wheel onto the motor shaft and mark a line on the wheel, you should be able to check this out precisely. The motor does not have any actual mechanical detents; this effect is due to one of the iron cores lining up with one of the magnets, a phenomenon described in my book *Ink Sandwiches, Electric Worms* (Experiment 21, "Motor Dice"). The most common motor of this type has a cylinder shape with two opposing flats around the circular cross section, typically about 15 mm across the flats, 19 mm in diameter.

If you can't easily buy such a motor, you can take apart a few discarded motorized toys until you find one. There are other motors in which the spacing

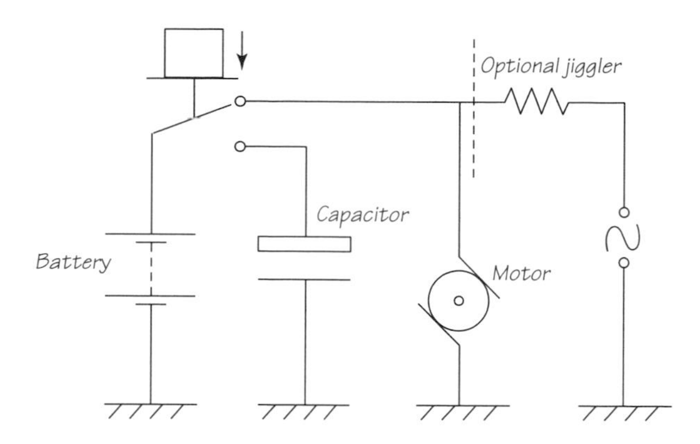

and small number of magnets and cores in the armature mean that they also have positive location angles around the axis. The small motor I recommend, however, has only six stable points; because these six points are basically equivalent in the motors I have, they are particularly suitable.

First you must set up the stepper motor circuit: the motor is arranged to be driven from the capacitor, not from the battery. Each time you operate the switch up and then down, the motor should jump to its next stable position. With each cycle of the switch up and down, the capacitor is first charged up, then discharged through the motor, causing it to hop along by one pole. I used pulses of 6 V from a 2,000 microfarad (μF) capacitor. With these values, I found that I could step six times per revolution, with a rate of 2 Hz. What seems to matter is the energy in the pulse provided. You could use a smaller capacitor, for example 470 μF, but charge it up to a larger voltage of about 12 V.

You can use a simple toggle changeover switch, but it will limit the speed of operation. If you can find a changeover microswitch or a push-button changeover switch, you can run the motor at higher stepping speeds.

Occasionally the 50-cent stepper will jam in a position between its stable positions. This situation becomes clearer if you put a small, lightweight wheel on the motor shaft, as suggested, with an index mark—a large arrow or something —so that you can clearly see whether the motor is stopping between positions. Friction from the commutator is probably mainly responsible for a motor's stopping between stable positions, but the externally connected load and any gearwheels connecting the motor to that load may also have some effect.

A separate "jiggler" current supply can be connected to deal with a sticking problem as follows: Use an AC transformer connected to the domestic electricity to supply low voltage AC. The AC low-voltage value can be anything from 3 to 30 V, since you should connect it via a large resistor to the motor. Choose the resistor value to yield a suitably low jiggler current: try a 500 or 1,000 ohm resistor to start with, or otherwise select a resistor that yields a current of around 10 mA.

How It Works

If you have access to an oscilloscope, you can see what is happening. Connect the motor to a battery and then connect the oscilloscope to it: you will see sharp, high spikes and longer, lower lumps in the voltage as the motor draws and then stops drawing current while the commutator rotates. Put a low resistance like 1 ohm in the motor lead (use a high-current resistor like a 1-W or 10-W type so you don't burn the resistor out), and then you can precisely measure the current flowing. You should find sharp pulses coming from the capacitor and less spectacular spikes coming from the motor: we calculate below how wide the capacitor pulses should be, but something on the order of 5–20 milliseconds is probably what you will get with the recommended values below.

The sharp, high voltage spikes derive from the motor's armature coil and its iron core. The motor coil stores small amounts of energy in the magnetic field in the iron core, and this energy is released in sharp spikes of voltage when that field collapses (i.e., when the motor is disconnected by the commutator). This is why electric motors are noisy, electrically speaking, and radiate radio-wave noise that can be picked up on radio and television sets. Most commercial commutator motors, like the ones in your vacuum cleaner or electric drill, use capacitors to suppress this effect.

HOW A COMMERCIAL STEPPER MOTOR WORKS

There are many different designs, but one of the most common and simple stepper motors has complex drive electronics powering four stator coils. These are stationed at 90-degree intervals around a moving rotor with six powerful permanent magnet poles. The rotor is a short, fat cylinder with three north poles at 120-degree intervals at one end of the cylinder and three south poles at 120 degrees, displaced by 60 degrees relative to the north poles. The system has twelve basic positions or steps that are 30 degrees apart, although it can be "half-stepped" to 15-degree intervals. (The illustration shows the permanent magnet poles as cut-out teeth, although in practice the rotor may simply be a cylinder, with the "teeth" existing only as a change of polarity.)

Essentially, the stators are fed with pulses to form what is effectively a moving magnetic wave, which the rotor follows. In normal mode, opposite

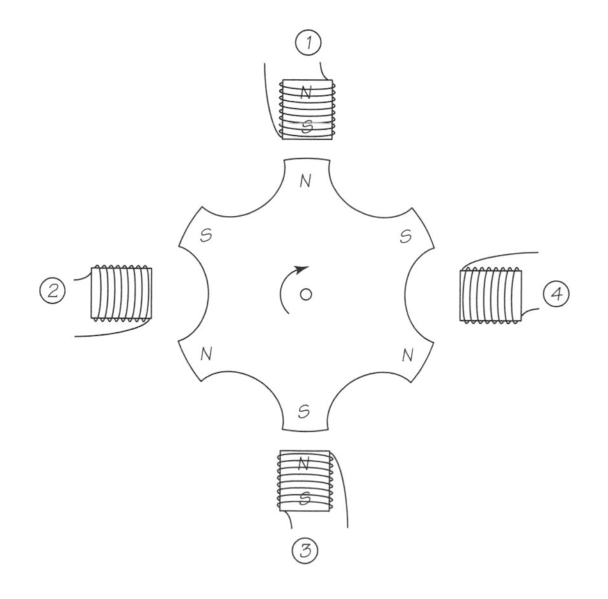

stators are switched to opposite polarities, whereas in half-step mode there may be three stators with the same polarity. The sequencing follows the following patterns:

Full-Step Mode

Substep	Poles	Poles
	1/3	2/4
1	N/S	N/S
2	N/S	S/N
3	S/N	S/N
4	S/N	N/S

Half-Step Mode

Substep	Poles	Poles
	1/3	2/4
1	N/S	N/S
2	N/S	-0-
3	N/S	S/N
4	-0-	S/N
5	S/N	S/N
6	S/N	-0-
7	S/N	N/S
8	-0-	N/S

Another design of stepper motor uses a similar drive electronics but achieves a much higher number of steps per rotation by using stators with four sets of poles spaced at a few degrees. The phased drive of the stator coils results in a set of, for example, twelve north poles that rotate around the axis, with twelve matching south poles following them. The stators are placed around the periphery of a simplified rotor, which has a large number of matching north and south poles in long, thin, longitudinal stripes. One I took apart recently had twelve north and twelve south poles around its periphery to match its stator coil poles. This motor therefore had forty-eight steps per rotation (7.5 degrees per step, or ninety-six half-steps at 3.75 degrees per step).

THE SCIENCE AND THE MATH

Imagine a vertical wave—a set of hills—with six peaks, along the top of which a ball can roll. The stable positions are the six "valleys" where the ball can roll, and the peaks must be overcome to move the ball on. Applying an impulse from the capacitor is equivalent to kicking the ball out of one trough over an adjacent peak, so that it goes down the other side and thus moves to the next stable position. The energy stored in the capacitor is transferred to the motor winding, which applies the impulse. This impulse must be more than the force-distance product that the magnet and the iron core impose—the energy level represented by the peak height. Each time the switch is operated, it first charges the capacitor with the impulse energy from the battery, then drops this charge into the motor. The amount of charge stored must be adjusted to turn the motor from one pole/iron-core pair to another.

The maximum speed of the 50-cent stepper motor is limited by the mechanical inertia of the rotating part—the rotor—and the time constant of the armature resistance/capacitor (RC) combination. Taking the latter first: with a 3-ohm motor and a 2000-μF capacitor, the time constant τ of the discharge into the motor coils is quite short:

$$\tau = RC = 6 \text{ milliseconds.}$$

With this kind of time constant, you could conceive of operating the 50-cent stepper motor at up to 100 pulses per second. However, the effective time constant is stretched by the commutator, which both adds resistance to the circuit—a resistance that is only partially seen in a static resistance measurement—and also reduces the effective current because it is delivering current only about 80 percent of the time.

The mechanical time constant is probably more significant but is more difficult to estimate. One might expect a number of mechanical effects to contribute: principally the inertia of the rotor, the friction due to the commutator, and the drag of the air inside the motor. We can posit some kind of equation for the rotor's mechanical behavior along the lines of

$$-I\, d\omega/dt = T_{friction} + T_{drag},$$

which is a rotational version of Newton's law for acceleration, where I is the inertia of the rotor, ω its angular velocity, and the T's are the friction and drag torques. This leads to

$$-I\, d\omega/dt = T_{friction} + (C_{vdrag}\omega) + (C_{tdrag}\omega^2),$$

where C_{drag} is the coefficient of drag, which varies with ω for viscous drag and ω^2 for turbulent drag. This kind of equation is hard work to solve, although a numerical solution is always possible by using finite increments in a spreadsheet, for example. It is clear that the general behavior is a slowing down—the forces all tend in the direction of slowing the rotor—and so the equation simply describes the curve of that deceleration.

The mechanical inertia I can be estimated by using the usual $I = \Sigma\, MR^2$ equation. The damping forces, however, are not as easy to estimate. You can get a good idea of the mechanical time constant through measurement: spin the motor up using a battery and then see how long it takes to slow down. You will probably observe a slowdown time that is a sizable fraction of a second, reflecting a time constant on the order of at least 100 milliseconds or more. This value is much greater than the electrical time constant and will dominate the problem of going faster: your stepper motor is probably not going to go faster than about ten pulses per second at best. This is why the stepper motor responds to the energy in a pulse rather than specifically to its voltage or the capacitor value: the electrical pulse behaves as a virtually instantaneous impulse of energy, from the point of view of the slowly reacting rotor.

And Finally . . . Stepping Up the Speed

Based on the theoretical discussion above, how fast can you make the 50-cent stepper motor go? If you double the size of the capacitor, does the motor jump on two steps at once, instead of one? Or do you need a much larger capacitor to get a double step? Are triple steps and more possible—or does the behavior of the motor become less predictable? Could you connect the motor to a transmission—or would the train of gearwheels stop the motor from stepping properly? With a 10:1 reduction transmission, for example, you will get 60 steps per revolution, just the thing for the second or minute hand of a clock. Curiously enough, if you look inside a quartz analog clock you will find a similar arrangement.

Does the commutator affect the results you get? It has spring-loaded contacts (the brushes) pressed against it, giving the device friction. The commutator is also not perfectly circular in cross section, because it is divided into sections—usually three of them in the case of these small motors.

Pushing String

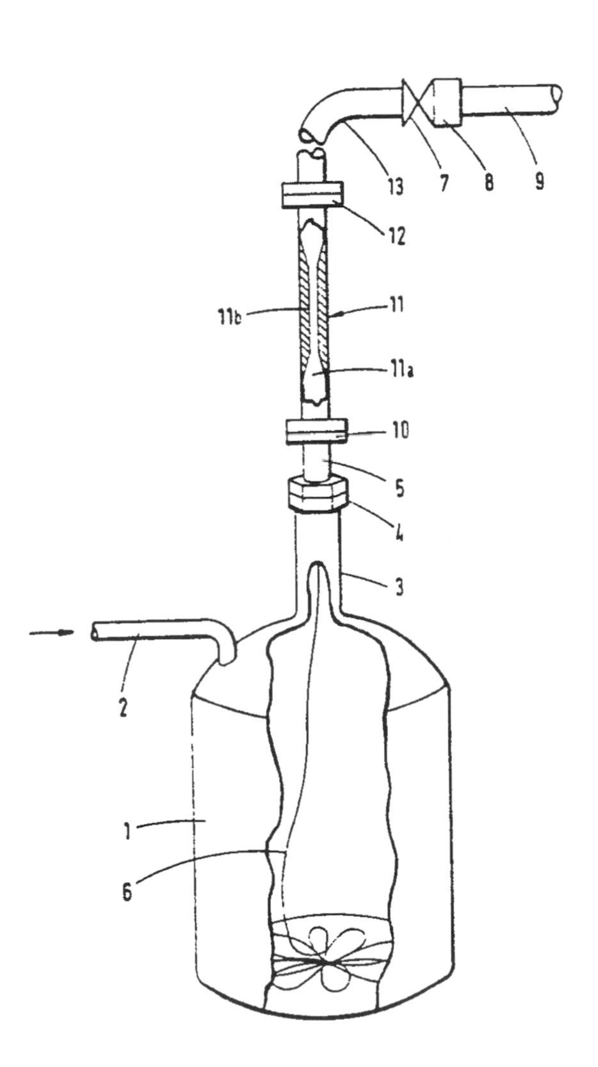

Pull a string, and it will follow wherever you wish. Push it, and it will go nowhere at all.

—Dwight D. Eisenhower (who often used this analogy of leadership)

I once organized a competition to see who could "push string"—propel it across the floor, using no more than hand power—while remaining behind a white line. The competition was based in a gymnasium, and I had expected that some competitors would devise ingenious gadgets to help, while others would use more skill but simpler devices. Throwing a large massive object, like a harpoon or a bowling ball, with string attached was banned; movement across the floor from a standing start was required.

The competition was good fun, although a bit chaotic. A couple of competitors came up with a kind of rocking stepladder device that I dubbed the string walker. The idea was that you pulled back the top of the walker gently, the string being connected so as to push out the walker's front legs a little as those legs came free from the ground. You then relaxed the string at the right moment and allowed the device to rock forward, whereupon a weak spring pulled the back legs back toward the front ones, the device having progressed forward a few millimeters in the process.

Although ingenious, the string walker was outrun by other devices. Amid the chaos a true winning strategy was not totally clear, but one that turned out to be surprisingly effective involved large food cans wrapped with string. The secret to getting the food cans to roll well was to pull hard and to accelerate them to a high speed in their first 3 feet of travel, while the cans were close enough that you were pulling more upward than backward. Skill was needed to keep the cans rolling in the right direction, against their tendency to twist and head off sideways. Some of the other ingenious ideas from the competition performed nearly as well as the cans, and might well have worked better than the food cans on a carpeted floor. We explore these rather more technical devices below.

String pushing, or at least something rather like it, is used for one or two practical purposes. We review an Inuit fishing system, for example, in the land jiggers project below. More important, the hairlike optical fibers that carry information in the modern world—the entire infrastructure of the Internet and telephones—are installed by a kind of string-pushing technique. Buried in our walls and in the ground under roads and sidewalks is plastic ductwork just a few millimeters in diameter, designed for containing optical fibers. Many of the fibers in these ducts are installed by "fiber blowing": the fiber is pushed through a narrow orifice while air is blown along the duct. Once a few meters of fiber have been pushed along, the minute drag on the fiber due to the air rushing past is enough, surprisingly, to pull the rest of the fiber along and to drag more fiber from the supply reel. Perhaps even more of a surprise is that the hundreds of meters of fibers don't immediately get hopelessly tangled.

Last we consider an alternative to the overhead electric cables, a system for the direct transmission of mechanical power using strings.

7 The Pull-me-to-push-me

"This, Doctor," said Chee-Chee, "is the pushmi-pullyu—
the rarest animal of the African jungles."

—Hugh Lofting, *The Story of Dr. Dolittle**

At least one kind of string-pushing device has been around for more than a hundred and fifty years: string-activated animals. The string-activated spider or mouse was a popular toy, as far back as the early nineteenth century and before. Still found occasionally in boxes of Christmas cookies, versions of this idea still persist, although rarely, in modern toys. One variation, for example, uses a rubbery spider propelled by a thread (a natural choice of animal to use thread activation for, of course, particularly if you use a silk thread). The theory explaining why these work, which needs a little effort to understand, is given below. Thread spider toys don't satisfy my requirement for a "true" string-pushing vehicle, however, since they require that you pull to some extent upward—pull horizontally and they just don't go. You can only propel them across the floor for a few feet, certainly not clear across a gymnasium floor.

Perhaps the simplest way of making a true string pusher is to use a vehicle that has a drum mounted parallel to its wheels and connected via a reduction transmission to the wheels. If the vehicle drum has a large amount of extremely thin string wound around it, and the drum is connected via a reduction transmission to some soft, gripping, rubber-tired wheels, then the system works unexpectedly well.

*For those who don't know: Dr. Dolittle's principal claim to fictional fame is his ability to talk to animals, from beetles to behemoths—and even a few plants. The pushmi-pullyu is an animal reminiscent of both a two-headed llama and a pantomime horse: with four legs, but with a head at both ends.

What You Need

- ❑ Drum for vehicle
- ❑ Erector or Meccano parts (maybe Lego Technic parts)
- ❑ Soft rubber wheels
- ❑ Strong thread such as button thread (large length, e.g., 30 m or 100 m)

Optionally

- ❑ Drum for thread (e.g., large thread reel)

Or

- ❑ Fishing reel and line

What You Do

You can build a pull-me-to-push-me in myriad different ways, but you do have to take care over some features. For example, you do need to ensure that the vehicle is either four-wheel or rear-wheel drive. With front-wheel-only drive, the vehicle tends to tip backward as you pull, reducing the grip, and it will tend to slip too easily. You also need to make sure that the vehicle has good grip on the surface over which it will travel.

Having built a suitable assembly qualitatively similar to that shown in the drawing, you now need to charge the drum with string. I found it easiest to push the vehicle along by hand, allowing the thread to run from the thread reel or the fishing reel onto the vehicle's drum.

With everything ready, you can now haul away at the connecting line. It is important to pull at the line evenly, not to snatch it, or you will pull the vehicle backward or, worse, off to one side. You should be able to get forward movement of the pull-me-to-push-me at any distance. You will also pull in whole heaps of string that will soon get tangled—so watch out! That is why I suggest that you use a "stationary" drum—just the thread reel itself if it is a good size or, even better, a fishing reel. That way you can conveniently reel in many yards of line, propelling the pull-me-to-push-me substantial distances.

With the right parts, you can try out different transmission gear ratios. With a small gear ratio and a smallish thread drum, the vehicle will be highly efficient in terms of what you might call the pull-me-to-push-me ratio (PPR), which can be defined as the ratio of forward distance S moved/length L of thread wound, i.e., PPR = S/L.

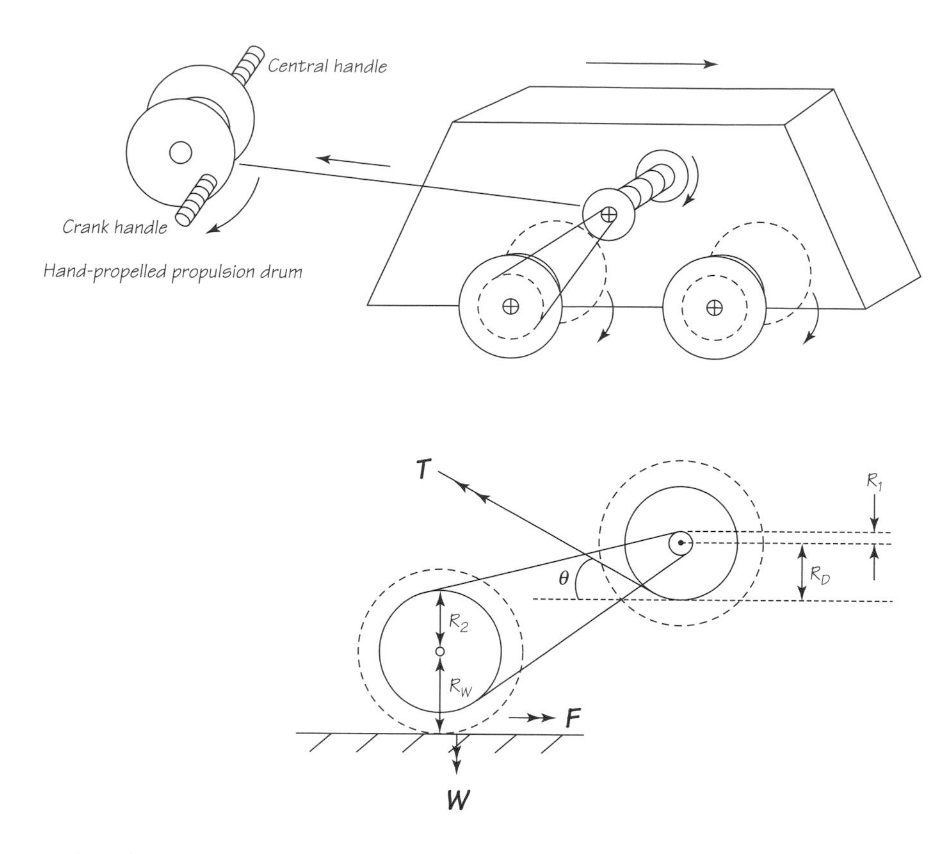

Overdo it, and you will find that your pull-me-to-push-me will become a pull-me-to-pull-me and go backward! With a lower gear ratio, the PPR will drop, but the vehicle will progress more reliably, albeit with greater consumption of thread. At low PPR values you will also be able to climb gentle upward gradients. The nature of the wheels also affects the steepness of the gradient that the pull-me-to-push-me will climb: wheels with soft rubber tires provide the maximum coefficient of friction and allow steeper slopes to be tackled.

THE SCIENCE AND THE MATH

The art of the pull-me-to-push-me is to choose the optimum compromise, steering between Scylla and Charybdis.* We have on one hand the Scylla of a transmission gear ratio that maximizes speed (and which needs a low reduction ratio), and on the other hand the Charybdis of a transmission that minimizes the resistance to being pulled back (and which favors a high reduction ratio).

The theory of the pull-me-to-push-me is straightforward for a horizontal pull on the string. If pull

*The vengeful mythological monsters who lurked respectively in the dangerous rocks and perilous whirlpools on either side of the Straits of Messina in Italy. Luckily, I saw no sign of them when I sailed through the straits a while back.

force = F_p, then with perfectly efficient gearing, drum diameter D, wheel diameter W, and a gear transmission ratio R, the tractive force F_t will be:

$$F_t = F_p DR/W.$$

By choosing W to be small, and R and D to be large, you can achieve any desired forward tractive force. However, choosing R and D to be large requires a huge length of string L_{string} relative to the distance moved S:

$$L_{string} = S(DR/W),$$

where the ratio is the same as the mechanical advantage we have seen above. We can express this another way by noting that, in the absence of inefficiencies in the transmission,

$$L_{string} = S/PPR$$

where the PPR, the pull-me-to-push-me ratio, is defined as above. The mechanical advantage ratio must be greater than 1, since

$$F_p < F_t.$$

So $F_p < F_p DR/W$, which is true provided that $(DR/W) > 1$.

Since the vehicle in theory requires only a very small forward tractive force to make it move with a very small acceleration, it is sufficient for DR/W to exceed 1 only by a tiny margin. In practice, it is necessary for DR/W to be appreciably larger than 1.

This is why the spider toy previously mentioned won't work with the string horizontal: the drum would have to be bigger in diameter than the wheel it operates, and this is clearly impossible. You could devise such a system using rails: the railroad wheels could be smaller than the drum driving them if the sleepers all had deep cutouts to allow a large-diameter drum to pass, one whose bottom half hung between and below the rails.

Achieving tractive effort requires that the vehicle does not slip. Slipping will not happen provided that the wheels are not simply dragged backward by a sudden pull on the string:

$$(F_t + F_p) < \mu Mg,$$

where μ is the coefficient of friction, M is the mass of the vehicle, and g is gravity. This requirement will always be satisfied if the machine has $(DR/W) > 1$, provided that the transmission system is perfectly efficient and that little acceleration is required. If, however, the pulley system has losses, or if the string is jerked backward suddenly, requiring a sudden acceleration, then it is certainly possible to cause the machine to slip backward.

This analysis assumes that the vehicle either has four-wheel drive or is just about to tip over backward, and hence has all its weight on its driving (rear) axle. It also assumes that there is no gradient. If we add a small gradient slope of an upward angle θ, then the equations change somewhat. There must now be a definite positive forward force urging the vehicle on:

$$F_t - F_p > Mg \tan\theta$$

$$F_p(DR/W + 1) > Mg \tan\theta$$

$$F_p > Mg \tan\theta \,/\, (DR/W - 1),$$

which clearly demonstrates that DR/W must be greater than 1 and PPR less than 1; otherwise F_p values will grow to the point where frictional grip μMg is exceeded. These equations also show that large DR/W values will reduce the impact of slopes.

If $F_t > \mu Mg$, then slipping will occur; in other words, $F_p \, DR/W < \mu Mg$ is the condition for forward progress, or

$$F_p < \mu Mg/(DR/W).$$

But we already have a condition to climb the slope from above, so we can write what we might dub the pull-me-to-push-me condition for F_p:

$$\mu Mg/(DR/W) > F_p > Mg \tan\theta \,/\, (DR/W - 1).$$

So we must select a value of slope such that there is an operable band for F_p. In other words, we must ensure that

$\mu Mg/(DR/W) > Mg \tan\theta/(DR/W - 1)$

$\tan\theta < \mu(1 - W/RD)$

i.e., $\tan\theta < \mu(1 - \text{PPR})$.

So with values of the PPR that are large (close to 1), the slopes that can be climbed are close to zero. Inspecting this further, you can see that with a very low PPR, the "1" in the expression can be neglected, which means that a slope can be climbed up to the limiting value of any vehicle on a slope, thus:

$\tan\theta < \mu$.

For a typical frictional value (such as brass on brass) of 0.35, this gives $\theta_{max} = 20$ degrees. Here, however, we use high-friction soft rubber wheels, which should give us a μ value closer to 1, yielding $\theta_{max} = 45$ degrees.

Simpler pull-me-to-push-mes

I previously mentioned a crude pull-me-to-push-me made from a simple drum with thread wound around it (see the diagram). Here you get forward motion if pulling force F_p produces a tractive force F_t that is greater than the horizontal component $F_p \cos\theta_p$ of that pull:

$F_t > F_p \cos\theta_p$.

But F_t is simply given by noting that

$F_t \sim F_p(D/W)$,

where D and W are the diameters of the drum and the wheel, respectively. Combining these we have:

$F_p \cos\theta_p > F_p(D/W)$.

So $\cos\theta_p > D/W$

and $\theta_p > \cos^{-1}(D/W)$.

For example, for $D/W = 0.95$, $\theta_p > 18$ degrees; for $D/W = 0.9$, $\theta_p > 26$ degrees; and for $D/W = 0.5$, $\theta_p > 60$ degrees. So it is important to ensure that the drum is nearly as big as the wheel.

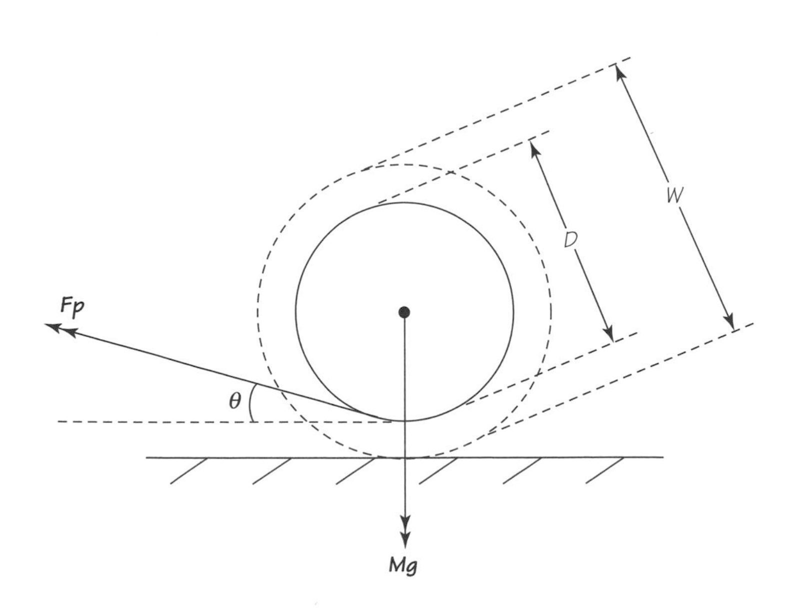

And Finally . . . the Steerable Panjandrum

It should be possible to make a pulley-operated version of the cable-drum vehicle. You would need a long and variable-length idler loop. This is the same problem that faced early computer engineers in the 1960s, when they needed to use magnetic tape loops of large lengths. The tape loop drives of the IBM, CDC, and ICL computers of that era usually employed long, vertical, hollow, rectangular prisms of clear plastic, filled with multiple S-shaped loops of tape hundreds of feet long; tape was added at the top and withdrawn from the bottom. Maybe a similar idea might work here?

Perhaps an even bigger challenge would be a steerable pull-me-to-push-me. There have been devices that have dragged two steering cables along after them. The Panjandrum was a device tried by the British military in the 1940s. It was a 10-foot-diameter drum bearing explosives, which would roll up on enemy beaches ahead of marines making a landing. The device, driven by small rockets on its perimeter, pulled small steel cables, one on each end of its central axle, along behind it. One version of the Panjandrum was to some extent steerable, using drag brakes on the two winches that unreeled as the device rolled ahead, as Gerald Pawle described in *The Secret War.* Surely an inversion of these principles should be possible, whereby a pull-me-to-push-me employing two separate drive systems, with motive power from two separate fishing reels, could be steered by differential winding in the cables onto the two reels.

REFERENCES

Downie, Neil A. *Ink Sandwiches, Electric Worms, and 37 Other Experiments for Saturday Science.* Baltimore: Johns Hopkins University Press, 2003.

Pawle, Gerald. *The Secret War, 1939–1945.* London: Harrap, 1956.

8 Land Jiggers

The Land Ironclads

—H. G. Wells, title of his book predicting military armored vehicles

H. G. Wells anticipated that technology in the early nineteenth century would soon allow the extension of the armored-battleship principle to land vehicles. He was not the only one to make such predictions, which probably started even before Leonardo da Vinci over four hundred years previously, but Wells anticipated the essentials of what became the modern battle tank better than anyone else. In this project, we make a similar but rather less far-reaching prediction that we can transfer another idea from its seaborne origins to the land: a device called the "jigger."

String-pushing technology has been serving humanity for at least a hundred years in a most surprising environment. Frozen lakes in Alaska and the northern provinces of Canada are teeming with fish. For much of the year, though, they are covered by several feet of ice. You might think this poses a problem for fishing—and you would be right. But the Inuit have devised a peculiar, almost unbelievable way of reaching those fish.

Inuit anglers have long used a device called the jigger, which crawls along underneath the ice, towing a thin rope behind. An Inuit first digs a hole through the ice and then lowers the jigger through the hole. The jigger, made of a relatively low-density wood, floats up until it lies underneath the ice, pressed up by its buoyancy against the ice's lower surface. Contrary to expectation, perhaps, the lower surface of a typical sheet of ice is relatively smooth. By repeatedly pulling and relaxing the rope connected to the device, the jigger first pulls itself forward, sticking a sharp pin in the ice, and then pulls the pin out and moves forward. The cycle continues, with the jigger edging itself forward.

After the jigger has proceeded a few tens of yards, the angler paces the length indicated by the amount of rope pulled out and then polishes the surface of the lake. The polishing enables the Inuit to see through the thin layer of frost and snow on top of the ice. The ice, although up to 2 meters (6 feet) thick, is very clear, and with luck the Inuit can see the brightly colored jigger. He or she digs another hole and retrieves the jigger, tugging the rope along through, which pulls a fishing net after it. After a few hours, the fish that are entangled in the net can be hauled out and taken home for supper.

Adapting this extraordinary device, we can build an upside-down land-based equivalent of it—the land jigger—for pushing strings across floors.

What You Need

- ❑ Wood
- ❑ Erector set parts
- ❑ Pin
- ❑ Strong thread such as button thread
- ❑ Spring or elastic band

What You Do

The diagram indicates how the land jigger works. It pulls itself forward by pushing a pin into the ground beneath, the pin being propelled by a lever connected to a string. The wood body of the device is slotted to allow the lever to act centrally. The lever pulls backward a long way when you pull on the string but pushes the jigger forward a smaller way, while exerting a higher force. A weak spring or elastic band returns the lever to its starting position.

After constructing a device along the lines shown in the diagram, place it on a surface. Try to choose a surface that the pin can grip effectively, but one in which pulling the pin out after it has stuck in is relatively easy. You'll have to experiment with different surfaces and difference styles and angles of pin to optimize the device on each surface. Start with something like smooth concrete,

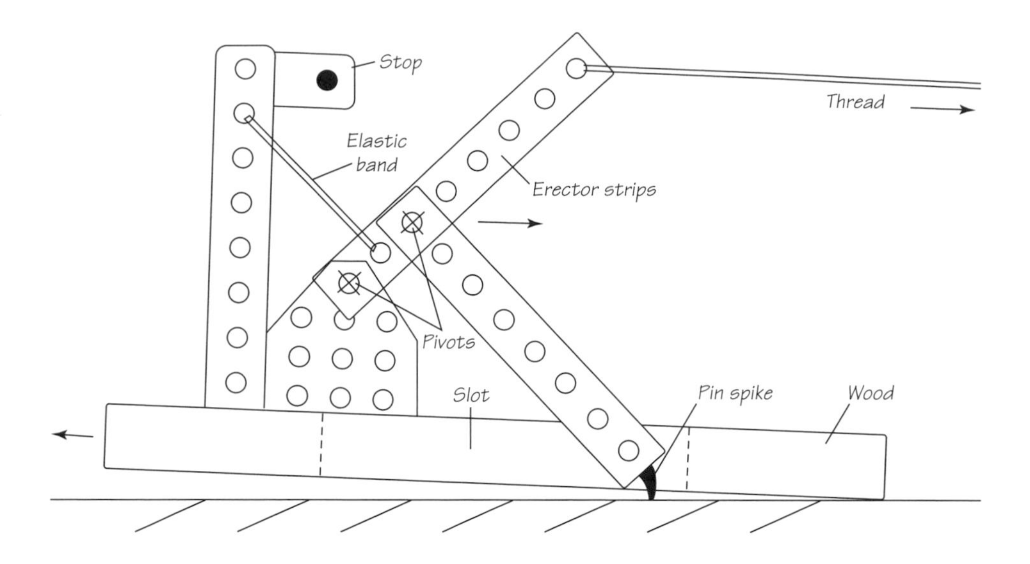

polyvinyl chloride (PVC—what used to be called linoleum), or wood. Try different pins: sharper versus more rounded at the end, thicker versus thinner, and so on. The precise length of each part of the device will vary the performance too, and not always in an obvious way (take a few minutes to read the math analysis below if you want some guidelines before experimenting).

How It Works

Pull on the string horizontally. You should find that the pin digs in, and then that further pulling draws the lever back and moves the land jigger body forward. Now relax the string. The return spring will pull the lever forward again, ready for another stroke, hopefully without slipping backward. The pin will pull out, since it does not have the component of the string pulling the pin down and digging it in, and the lever will then move forward, dragging the pin along the surface. The land jigger is now ready for another move forward.

The smoother the surface, the better your land jigger will crawl along, provided that the pin can dig in but still pull out smoothly. The Inuit jigger has the advantage of working with the ideal surface: ice. The force needed to pull a pin out of ice is low, even if the pin is pushed in hard. Also, the sliding friction of the Inuit jigger on ice, with only a fraction of its weight pushing upward as a normal force, is lower. For some surfaces you could try using a small pad of gripping rubber: this may work better than any kind of pin.

Once you have mastered the action of making the jigger go, you can try timed distance trials or a race. You should find that, although the device may run off course, it can be "steered" by pulling back toward the side opposite the course deviation, using an angle bigger than the deviation.

THE SCIENCE AND THE MATH

The land jigger relies on the leverage that its lever gives it on the ground. Although the lever (see diagram) is pulled backward with a force F_p, it pulls the machine forward with a tractive force F_t. This force is related to F_p, because the force F_p exerts a force $F_p(L/P)$ on the end of the link q. The component of this force along the link q is $F_p(L/P)\cos\theta$. And the component of this force F_t exerted on the gripping pin on the ground is

$$F_t = F_p(L/P)\cos^2\theta,$$

where L is the string-to-pivot distance and P is the arm-to-pivot distance. (It is not quite this good in practice, because of the undesirable effect of the return spring.) The vertical component of this force F_v is given by the similar expression:

$$F_v = F_p(L/P)\cos\theta \sin\theta.$$

But $F_t < \mu F_v$, otherwise slipping of the pin will occur, where μ is the coefficient of friction. This means that we have a first limit on the operation of the land jigger:

$$\tan\theta > 1/\mu.$$

For example, with a coefficient of friction for the pin of 0.5, then $\theta > 63$ degrees. With a coefficient of 0.2, a value more typical of a slippery surface, then θ must be > 78 degrees. This is clearly a limitation on the jigger, which would have a very limited

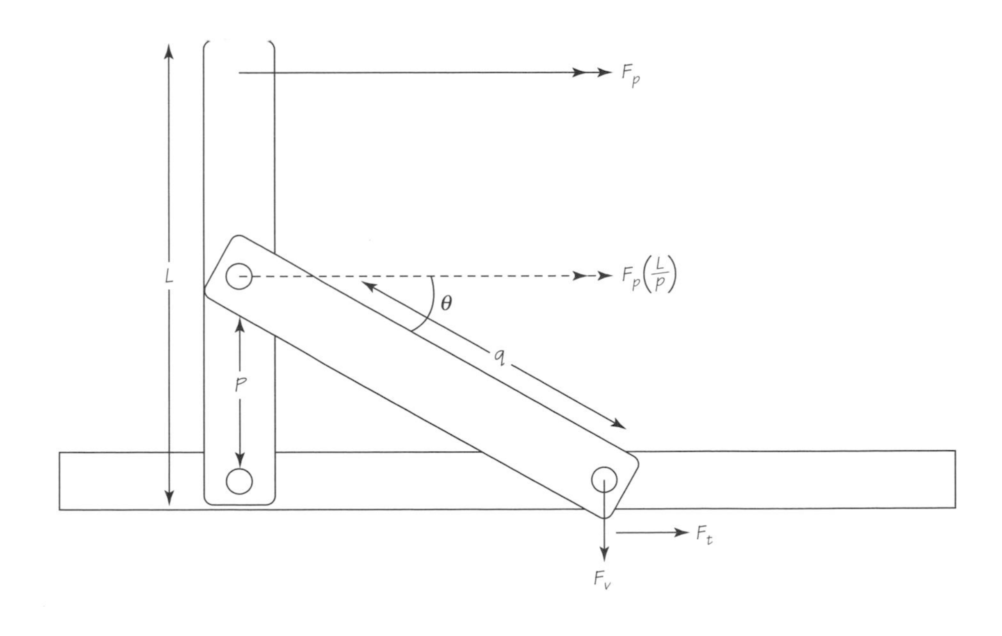

stroke for each cycle if it had to use such a large angle for θ. In practice, the coefficient of friction is effectively much larger because the pin digs into the substrate, and angles like 45 degrees are usable.

Another limit on the land jigger ensures that the device can overcome the friction due to its own weight. With a weight of Mg, the device must be pushed along with an F_t of greater than μMg. But the net F_t is actually F_t less the pulling-backward force on the string of F_p:

$$F_t - F_p > \mu Mg$$

$$F_p ([L/P]\cos^2\theta - 1) > \mu Mg.$$

This will not be possible unless $(L/P)\cos^2\theta$ is > 1, in other words,

$$L/P > 1/\cos^2\theta.$$

Taking typical values again, suppose $\theta = 70$ degrees; then L/P must be > 10. But with the pin digging in, much smaller angles θ become usable. If we can use 45 degrees, then $L/P > 2$ will be workable. But what is the effect of the return spring or elastic band?

And Finally . . . the Toothbrush Shuffler

Can the land jigger attain greater speeds? With a bigger lever, the land jigger could move a longer distance per stroke. With the same size of lever, you could speed up the device by rhythmically pulling the string at a higher frequency. But what is the maximum size of lever? Can it be longer than the vehicle itself? What is the maximum frequency of string pulling? Is it limited by the moment of inertia of the lever? Would a lightweight carbon fiber lever or a lightweight string—a Kevlar thread, perhaps—be better? Or is the lever pulling's speed limited, like a pendulum, simply by the lever's length? If the lever is a true pendulum, then surely the longest lever possible is preferred. Although the natural frequency of a pendulum decreases with the square root of length, the length of stroke increases linearly with that length, so overall you gain by a factor of square root of length.

During the competition among string-pushing devices that I organized, several kids attempted to make a device related to the land jigger that employed a specially oriented toothbrush. This toothbrush shuffler employed a similar string pulling on a lever in a rhythmic way, but it used a trimmed toothbrush instead of the land jigger's pin. The examples I saw did not work well, but can you do better? Could a toothbrush shuffler be made to function better than a land jigger?

REFERENCE

Downie, Neil A. "Dougall or Vibrocraft." In *Vacuum Bazookas, Electric Rainbow Jelly, and 27 Other Saturday Science Projects*, 151–56. Princeton, N.J.: Princeton University Press, 2001.

9 Power Strings

> So we descend from the great and esoteric heights [of Maxwell's equations] and turn to the relatively low-level subject of electrical circuits. We will see, however, that even such a mundane subject, when looked at in sufficient detail, can contain great complications.
>
> —Richard P. Feynman, in *The Feynman Lectures on Physics*

We are used to the idea that power can be transmitted over large distances by means of overhead electric cables, supported by steel pylons. Aluminum wires that thread their way across all the developed world are a familiar part of the landscape. (Overhead wires are made of aluminum because it is both cheap and technically superior. It is nearly as good a conductor as copper, but it has a lower density, so its conductivity-to-density ratio is twice that of copper. The wires have a core of high-tensile steel, which conducts little current but provides most of the tensile strength.)

Only a hundred years ago there was no such large-scale power transmission. Power had to be generated where it was needed by a variety of engines. Only local power distribution was possible. Solid iron shafts in the roofs of large factories once carried power in rotary form over a hundred yards from a central steam engine to subsidiary machines such as power looms or drills.

In early industrialized cities, such as Manchester, U.K., high-pressure pipes conveyed hydraulic power from a central station to different factories over longer distances, perhaps a mile or two. Standard belts between two simple pulleys cannot be used over more than a few dozen yards. However, power could be transmitted for several miles by hauling steel-stranded cables over many intermediate pulleys. A vintage multiple-pulley-guided cable system powers the famous San Francisco cable cars. This system has a constantly moving cable running along pulleys a couple of feet below the street surface: the streetcars access the cable via a slot.

Here we explore the possibility of transmitting mechanical power via rotating pairs of tensioned cables. You may be surprised how efficient such a system could be, although it might be more usable on a planet other than Earth, for reasons that will become clear.

What You Need

- ❑ Two motors (e.g., small 1.5–3-V types, or perhaps one standard type, one high-impedance type)
- ❑ Battery/battery box
- ❑ Multimeter
- ❑ Nylon fishing line, 0.24 mm in diameter (3.8 kg breaking strain rating)
- ❑ Wood
- ❑ Glue
- ❑ Wire

What You Do

This system transmits rotational power from one pylon-mounted motor to another pylon-mounted motor that is used as a generator and thus receives the power. The small wooden rotor on the motor pulls on a pair of thin strings to make the generator turn. The two motors must be as close to coaxial as possible. If you fix the pylon positions and then glue the strings on, you can get more or less the same tension on the two strings.

Complete the setup as shown in the diagram, with a little slack but not so little that you can easily force the transmitter rotor around on its axis more than 90 degrees with respect to the receiver rotor. The cross-brace strings are optional (see below). Now switch on with the lowest available voltage. The system should begin to revolve freely and fairly fast, even with low power applied. Correctly adjusted, the fishing line's "orbit" will belly out to form a smooth, ellipsoidal

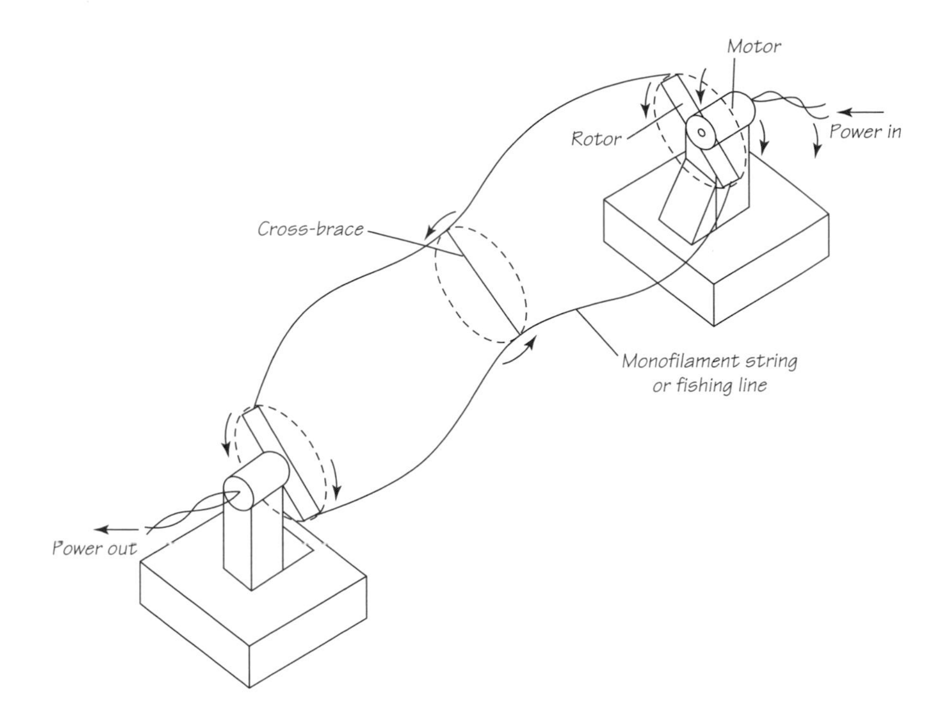

shape in the air, like an elongated version of an American or rugby football. If you've left too much slack in the strings, they will twist into a waist in the middle and fail to work. If you've stretched the system out with too little slack, then you may find that the motor labors because of the large axial force on its bearings. Check the voltage induced in the receiver motor: I got 0.6–0.8 volts in the system I set up, for 3 V or 300 mA on the input motor.

If you can fit, at the receiver end, a small "solar motor" (meant to run on the power from a low-power solar panel), you can transmit a higher voltage. These motors have a larger number of turns of wire on the armature, so at the same rpm they yield a handy 6 V or so.

Now try optimizing the system. Varying the tension in the strings is probably the first thing to try, most easily by changing the distance between the transmitter and receiver. See how much power you can apply to the transmitter. At higher power settings, you will hear a wailing sound that grows with the power applied. This is acoustic-frequency oscillation, caused by the air flowing past the fishing line.

The system can exert a certain maximum torque, limited by the twist problem alluded to earlier: with too much torque, the strings twist together; once twisted, they transmit little torque. The maximum torque that can be transmitted increases once the system is rotating, however, because the centripetal force on the strings opposes the tendency to twist, and a larger torque can be tolerated.

The force of gravity can be essentially ignored here, as can any stiffness in the strings: centripetal and drag forces dominate the motion of the thin strings you are using. The exception would come only with very long distances between the pylons, when a shallow version of the traditional catenary (suspension bridge) curve would begin to be superimposed on the ellipsoidal shape as gravity forces became more important.

The basic system can be enormously boosted in performance by the addition of simple cross-bracing strings. First, a cross-brace reduces the height of the pylons you need by decreasing the volume occupied by the envelope of the cable orbit. Second, a cross-brace reduces drag losses by lowering both the distance traveled by the strings in each revolution in their orbit and the speed of the strings in that orbit. With the aid of cross-bracing, the power strings can be extended to the maximum length possible, indicated by the straight-line calculation of maximum power transmission (see the math section below). I suggest you try cross-braces that are simply pieces of the same string you have used for the power strings themselves, carefully knotted or glued so that they don't weaken or bend the power strings. However, if a rigid cross-brace is used, it provides resistance to twisting, and the interpylon distance can be increased considerably above the normal limit—perhaps as much as double.

This system can thus conduct power over a remarkably large distance between just two points. For even longer distances, the same system can still be used: pylon supports, though, can be included to provide intermediate stations for the system. This is more complicated than simply adding cross-braces: pylons must include two bars and a bearing, and there are more string tensions to adjust. But pylons can be used to double or triple the usable power-transmission distance, since in effect each provides a new transmitter and receiver pair.

For convenience in demonstrating a power strings system and for measuring its performance, we use both electrical input and electrical output. In a larger-scale, practical power strings energy distribution system, the input would likely be a "prime mover" engine such as the steam turbine of a power station. This prime mover could be directly connected to the strings system. No alternator would be needed to convert mechanical power to electricity, as in an electrical distribution system. And receiving stations might well use the mechanical energy directly in rotational form, just as belt systems did in old factories, where they powered anything from corn grinding to weaving.

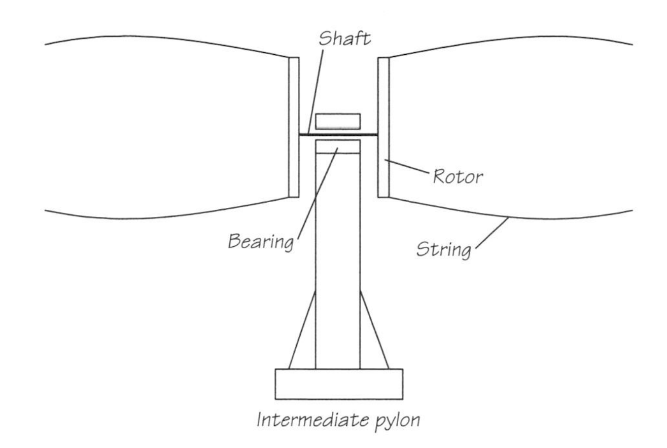

Try maximizing the amount of power you can obtain through the system. In the presence of air, you would expect the drag force to vary with the square of the speed at which the string goes through the air—that is, v^2 (see below)—and thus the power losses vary as v^3 while the power put in is directly proportional to v. This relationship means that you could expect to see a string speed v_{max}, at which the power transmitted is maximized.

THE SCIENCE AND THE MATH

In the case of straight-line cables between the transmitter and receiver stations, we can calculate the maximum possible power transmission by noting that power P transmitted is simply the product of force times speed, or, in this case, torque T times angular speed ω:

$$P = T\omega.$$

Now look at the schematic diagram showing the geometry of the system: if the torque T involves a force F_t at radius R, then the net torque of the two-string system is

$$T = 2F_tR.$$

The force needs to be calculated in two ways: whichever method gives the lower value of torque becomes the value of T used above. The first calculation requires that the cables cannot be tensioned beyond their yield (stretching) point, and then allows for the geometrical factors in the equation. The cables in my case were monofilament nylon fishing line, 0.24 microns in diameter. Now the tension F in the strings will be given by

$$F_t = F \sin\theta$$

so $F_t = T/(2R) = F \sin\theta$, and thus $T = 2RF \sin\theta$.

But $\sin\theta = 2R \sin\varphi/L$,

so $T = 4\ R^2F \sin\varphi/L$.

This formulation gives the torque transmitted for a given maximum tension F in the strings. The value of this torque shrinks as the length L grows, but it

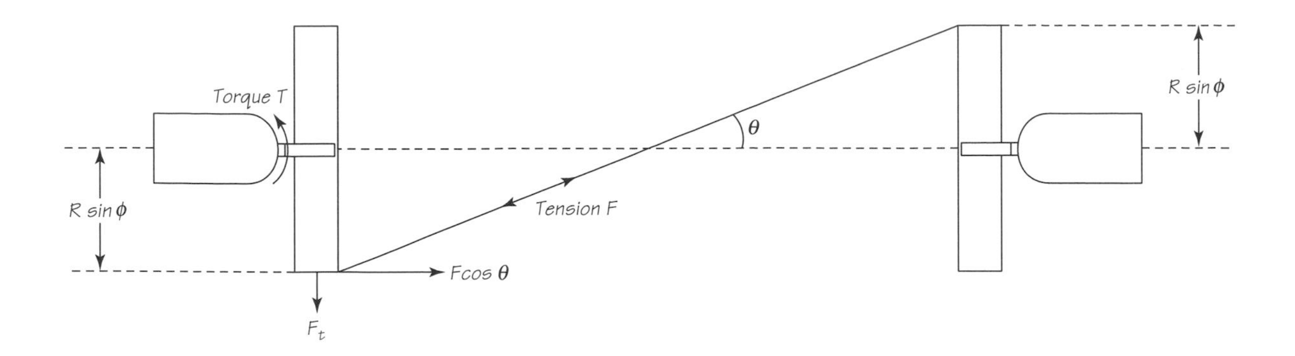

grows rapidly with R, indicating the need to maximize R if possible. The larger the "lag angle" φ, the larger the potential torque transmission too.

The second calculation checks that the cable system cannot get a twist. Twisting occurs if the cables are not tensioned sufficiently or are too elastic. This will happen when a transmitter exerts too much torque on the two cables. Once twisting has occurred, the amount of power that can be transmitted is sharply limited, while large axial forces will pull on the transmitter and receiver bearings. Typically, twisting will occur only in a static position or at low speed. Once a power strings system is running at a high enough speed, then centripetal forces will hold the strings out and reduce the tendency to form a twist. We can obtain an idea of when the strings might twist by noting that the extension of the string ΔL is given by

$\Delta L = L(1 - \cos\theta)$

which we can approximate by noting that $\cos\theta \sim 1 - \theta^2/2$,

so $\Delta L \sim L(1 - 1 - \theta^2/2)$, $\Delta L \sim L\theta^2/2$.

Now $\theta \sim R \sin\varphi/L$

so $\Delta L \sim R^2 \sin^2\varphi/2L$.

But the extension ΔL_s due to elasticity can be estimated using the definition of elastic constant E (Young's modulus):

E = Stress/Strain

$E = F/\pi r^2/(\Delta L_s/L)$.

So $\Delta Ls = LF/E\pi r^2$.

If $\Delta L_s < \Delta L$, then no twist should be possible. That is, if

$LF/E\pi r^2 < R^2 \sin^2\varphi/2L$,

$F < ER^2 \sin^2\varphi\, \pi r^2/2L^2$, and (from above) that
$T = 4R^2 F \sin\varphi/L$,

$T < 2ER^4 \sin^3\varphi\, \pi r^2/L^3$.

This calculation demonstrates that to maximize the torque transmitted, we must especially maximize the transmitter arm radius R and minimize the length of the interpylon distance L, while ensuring that we operate with a lag angle φ fairly close to the maximum 45 degrees. The actual value of the Young's modulus, perhaps surprisingly, is not quite so important, although carbon fiber would clearly be superior to rubber elastic-band material.

Drag Effects: Power Strings on Mars and the Moon

Thus far, we have tacitly assumed that our system is working in a vacuum environment. This might be a valid assumption for power strings on the moon or on Mars, but not on a planet such as Earth. Even at relatively low speeds, drag losses will be high—they actually dominate all other losses at higher speeds.

When you suspect that fluid motion—gas or liquid flow—might not involve turbulence, but also might not be pure "viscous" flow, you should compute the Reynolds number, Re. Re is given by

$$\mathrm{Re} = \rho v D/\mu$$

and expresses the ratio of inertial to viscous forces. Here ρ is the air density, v its speed, D the diameter or other characteristic length of the object/system, and μ is the air viscosity. When a ball bearing moves slowly through treacle or molasses, viscous forces (the μ factor) dominate, and flow will be smooth and viscous in nature. The Reynolds number will be small, less than 1. When a 500-ton jet plane moves through the air a hundred meters per second, air flow is largely turbulent, and the plane leaves behind a whole vortex field of rotating air masses. Viscosity effects are then completely negligible, and we would expect the Reynolds number to be in the 100 million or billion range.

The drag force on a cylindrical cable in a simple flow of air obeys a straightforward mathematical approximation. The actual turbulent nature of the flow, including effects such as the wailing noise (see below), is such that there is no exact formula analogous to Stokes's formula for the drag on a sphere in a viscous medium. However, between Reynolds numbers of 100 and 100,000, the standard drag formula

$$D = \tfrac{1}{2}\rho C_d A v^2$$

applies, where A is the area that the cable projects in the direction of flow, and C_d is the coefficient of drag, which for a cylinder has a value of 1–1.2. Allowing a larger variation in C_d from 0.5 to 2, the same formula can be applied over an even larger range, for Re between 10 and 1,000,000. The formula makes no mention of viscosity, because viscosity forces are negligible compared to those that result from inertia, pressure drag, and turbulence. Pressure drag is "pushing-the-air-out-of-the-way" drag, while turbulent drag arises from the energy needed to create the vortex motions that the string leaves behind it, which create the wailing noise.

In the case of power strings, we can calculate the power loss roughly by assuming that the strings follow a cylindrical orbit, and that the drag directly subtracts from the applied torque force. So you would expect the total power transmitted to be simply

$$P_{out} = P_{in} - P_{drag} = av - bv^3,$$

which is a curve that has a maximum value when $v = \sqrt{(a/3b)}$, as you can see by looking at where the differential $dP/dv = a - 3bv^2 = 0$. The cubic curve also has zeroes at $v = 0$ and $v = \sqrt{(a/b)}$: the latter zero is where the input power is precisely balanced by the drag force, and the power strings become simply a curious way of stirring the air, transmitting precisely no power at all! Now let us evaluate the constants a and b:

$$P_{in} = T\omega = 2F_t R\omega$$

$$P_{drag} = 2D\omega\tfrac{1}{2}\rho C_d A v^2 \text{ (for two strings).}$$

But $v = R\omega$, so

$$P_{drag} = D\omega^3\rho C_d A R^2.$$

So we have a maximum when $\omega = \sqrt{(2F_t R/3\rho C_d A R^2)}$, giving us the power strings equation:

$$\omega_{max} = \sqrt{(2F_t/3\rho C_d A R)}.$$

Hence the maximum rpm increases at higher values of the torque force F_t, and decreases with denser gas and the area of the string A, which is proportional to its length × diameter. It should not be a surprise that ω_{max} decreases with increasing R, since the drag force will clearly increase with increased orbit diameter.

Wailing Strings

The wailing of wires in telephone poles is well known to readers who live in or have visited places that have both strong winds and wires on poles. It has nothing to do with the well-known phenomenon of the vibration of strings in musical instruments. It arises rather from the formation of turbulence of a very well-defined kind in the wake of a long cylindrical object in a fluid stream. A "vortex street" is formed, in which vortexes are produced by the wire, one to the left and the next to the right, in a perfectly regular way. For most Reynolds numbers, for example, from 100 to 10 million, the frequency f of the wailing is given approximately by the simple equation:

$$f = vS/d$$

where v is the air speed and S is the Strouhal number, which is about 0.2. Here, for example, if we have a 600-rpm string on a 50-mm rotation radius, then we have stream velocity of:

$$v = 2\pi(50/1000) \cdot 10 = 3.14 \text{ ms}^{-1}.$$

With our string of diameter 0.24 mm we should expect a wailing frequency of:

$$f = 1000 \cdot 3.14 \cdot 0.2/0.24 = 2.6 \text{ kHz},$$

which accords well with what we heard from the power strings we tried.

And Finally . . . Transformer Stations

Clearly the power strings system could work better with thinner string. Maybe stainless steel line, sold in fishing tackle stores, would be good. Pylons obviously could conduct power strings around corners with the aid of gearwheels. However, the use of constant velocity (CV) joints or perhaps simply universal (Hooke) joints, as illustrated on the diagram, would allow the use of smaller angles in a chain of pylons. Perhaps less obviously, Hooke or CV joints could be used to permit long spans between pylons where gravitational sag—that catenary curve again—is a problem.

One of the beauties of electrical power transmission is that it is easy for power from a single large power station to be split up into small amounts to supply thousands or millions of individual users in homes and factories. Power strings' power can be split too, by dint of gearbox transmission units on pylons with two or more outputs in different directions to be fed to various locations. Differential units can be used to ensure that torque is divided evenly between two or more different users. (There could be problems if one user jams his or her power outlet. With a differential transmission gearbox, when one outlet is jammed, the other outlet or outlets all speed up. With two outlets, the unjammed outlet will run at twice the input shaft speed. Anyone who has gotten a car stuck in a muddy rut and seen the huge amount of mud that the nonjammed wheel flings out will understand this dynamic.)

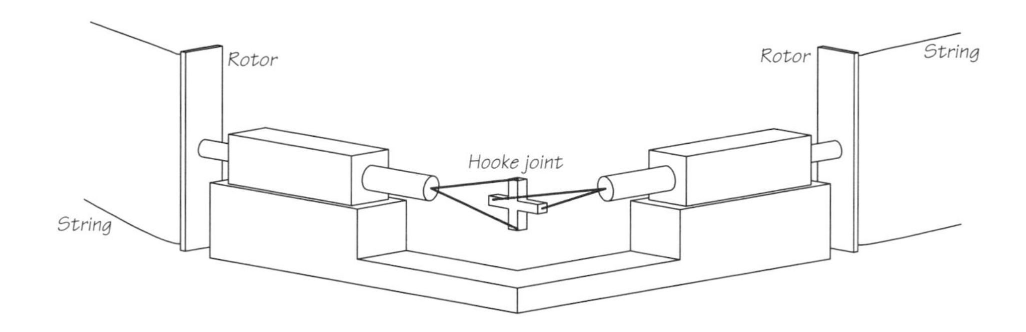

Power strings systems can also function as the equivalent of transformer stations. A set of power strings pylons can transmit large amounts of electricity at high speed and low torque (to applications that need low speed and high torque) by adding a "power strings transformer"—an ordinary gearwheel reduction transmission—at the receiver end.

Electrical power is used, of course, for all sorts of purposes. Two of the most common uses, heating and mechanical energy, can be readily duplicated by a power strings system. Heat requirements, for example, would be satisfied by use of a frictional brake device. It might be difficult, however, to supply illumination requirements via power strings: a flint bearing on an iron wheel would work as a light source, but it seems a touch primitive.

Would it be more effective to use more than two strings in each power link? Would three strings in a triangle be better, or four in a square, or even more? What about an infinite number of strings distributed around a wheel on the transmitter and receiver shafts? But wait a minute, I hear you ask: Isn't an infinite number of strings the equivalent of a cylinder, and isn't that a driveshaft system? Well, not quite, because driveshafts are rigid and have a problem with centripetal force, whereas an infinite strings drive would be flexible and, like the double-string power strings system, would use centripetal force to its advantage.

Peculiar Vehiculars

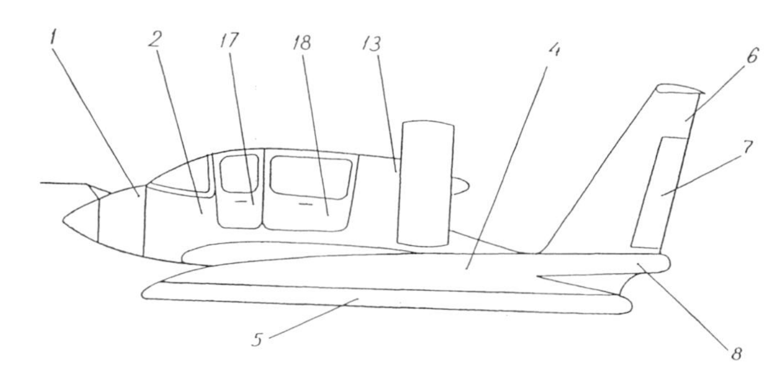

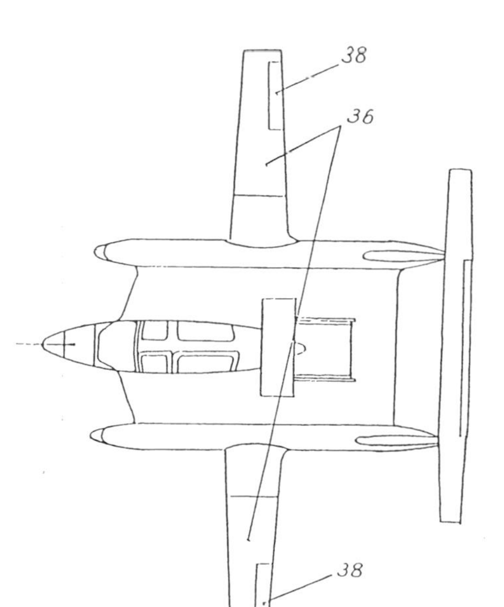

Фиг.4.

I think it's too bad when aviation movies depend for their excitement on plane wrecks and lost flyers and all that sort of thing. Perhaps that's good drama, but it certainly isn't modern aviation. . . . That's no more representative [of aviation] than a train wreck every half hour or so would be truly representative of rail transportation.

—Amelia Earhart, pioneer aviator, reported in *Screenland,* June 1933

In the early days of aviation, airplanes were mostly regarded as exotic and unreliable forms of transportation. In this section, we take a look at some other exotic and unreliable ways to get from A to B.

Travel is easy in the world of Newtonian physics—it is stopping and starting that require effort. Push off, says Newton's law, and you will sail on serenely until you need to stop, when you will have to push on something again. Newton's simple physics, however, only applies directly to motion in the vacuum of empty space. The force of gravity confines most earthly vehicles to move across a surface, encountering frictional forces. Aircraft, which avoid these surface friction forces, must contend with yet another formidable force: air drag.

Our first project, the ekranoplan, illustrates a maverick approach to reducing the large drag forces usually encountered by airplane wings. Normally airliners prefer to fly reasonably high in the earth's atmosphere, something over 10,000 m (30,000 feet), where air drag is appreciably lower. However, there can also be advantages in flying lower. If an airliner flies less than 3 m (10 feet) from the ground, drag is also reduced, quite dramatically.

The next two projects in this section, the slimemobile and the ball river bobsled, both explore novel ways to reduce the friction of air-driven ground vehicles, with dramatically different levels of efficiency in terms of maximum speed. (No prizes awarded for guessing which is faster.) We look finally at a vehicle whose entire modus operandi relies on friction.

10 Ekranoplan

> Science, freedom, beauty, adventure: what more could you ask of life? Aviation combined all the elements I loved. There was science in each curve of an airfoil, in each angle between strut and wire, in the gap of a spark plug or the color of the exhaust flame. There was freedom in the unlimited horizon, on the open fields where one landed. A pilot was surrounded by beauty of earth and sky.
>
> —Charles A. Lindbergh, *The Spirit of St. Louis*

A ram wing is a peculiar sort of aircraft that operates close to the ground, using wings rather than fans for lift. There is thus a sense in which the ram wing is to the airplane as the hovercraft is to the helicopter. Like an airplane, it must create a pressure difference by its forward motion: a lower pressure over its upper surface and/or a higher pressure underneath. You can illustrate the concept by simply taking a sheet of paper on a table and blowing at its edge. Do this right, and you can blow the paper 2 to 3 meters (6 to 10 feet) or more.

As it flies, the ram wing is supported by a pressurized zone of relatively static air beneath it. This dynamic is in contrast to that of a conventional wing, which relies on the reaction from a mass of air that has been set in motion downward. The pressurized air below a ram wing tries to flow away sideways, but, because the ground is so close beneath, it cannot escape sufficiently quickly to reduce the pressure; the remaining pressure is sufficient to support the aircraft. A ram wing is considerably more efficient than an ordinary wing in free air, other things being equal. Conventional winged aircraft can also benefit to a small extent in the same way if they fly along close to the ground, when they are described as operating in "ground effect." The lift/drag ratio of a ram wing depends upon its clearance from the ground, but if the ground clearance is small compared to the wingspan, the lift/drag ratio is much smaller than that of a wing flying at a height of a few wingspans. A wing flying low has more lift and less drag than a high-flying wing, but only if it is low enough to mow the grass.

The former Soviet Union developed one type of ram wing to a high order of efficiency. This particular aircraft, with a wing at the front and a high T-tail, was called an *ekranoplan* in Russian, *ekran* meaning screen or planar surface, and *plan* meaning plane or wing. I think of a ram wing as the generic term for a craft with one or more wings that flies within ground effect, with the ekranoplan as a particular subspecies. Soviet engineers tried a number of different designs with jet engines as well as propellers. The largest machines, nicknamed "Caspian Sea Monsters," were as big as airliners. Using eight jet engines, they could travel at enormous speeds approaching 200 miles per hour and carry tens of tons of payload. There were problems, however: these craft needed huge amounts of power to take off (or to unstick, as the process is picturesquely known), hence the set of eight engines. Once airborne a few feet above the sea, all but a pair of engines could be shut down. They were also rather unstable longitudinally and required great skill in piloting. The danger was that if an ekranoplan pitched up slightly at the nose, it might rear up uncontrollably.

In this project, we devise an ekranoplan that is fairly stable and easy to operate, so that the characteristics of this unusual form of aircraft can be tested.

What You Need

- ❑ Large flat area (e.g., tennis court, gymnasium, segment of a parking lot)
- ❑ Sheet of expanded polystyrene for the wing, measuring 50–70 mm (2–3 inches) thick by 500 mm (18 inches) deep by 600 mm (24 inches) wide (this material is sometimes called Styrofoam and sold as wall insulation for buildings)
- ❑ 2 pieces of plywood for constructing airfoil shape, 500 × 75 mm (18 × 3 inches) in 4- or 6-mm (3/16- or 1/4-inch) thickness
- ❑ Sheet of balsa wood, 6 mm (1/4 inch) thick, for fuselage and fin
- ❑ Sheet of balsa wood, 3 mm (1/8 inch) thick, for tail and wing/fuselage joint
- ❑ Motor (e.g., model-car motor, 540 type)

- ❑ Propeller from model shop, 200 mm (8 inches) in diameter by 100 mm (4 inches) pitch
- ❑ Bush with which to mount propeller on motor
- ❑ Wheels (I used parts salvaged from large toys)
- ❑ NiCd battery, 7.2-V 6 sub-C NiCd cells type, as used with radio-controlled model cars
- ❑ Battery charger
- ❑ Strong fishing line
- ❑ Small eyelet (to attach line to fuselage of aircraft)
- ❑ Long, 100–150-mm (4–6-inch) eyelet screw (sometimes called a "vine eye," since these are used to fix wires a little above a wall for climbing plants to grip)
- ❑ Bolt and washers (for eyelet screw to swing on)
- ❑ Wood for central post
- ❑ Wooden strip, 18 × 36 mm (3/4 × 1 1/2 inches), about 2 m (6 feet) long
- ❑ String
- ❑ NiCr wire, 0.56 mm in diameter (24 gauge)
- ❑ 6-V DC current supply (e.g., from battery or battery charger)
- ❑ Water-based contact glue
- ❑ Hot-melt glue

How to Build an Ekranoplan

The daunting list of parts above may discourage you. Don't despair! This project is simpler than any kind of radio-control model airplane because it does not, at least in its simplest form, have any form of control. It is also much more likely to fly successfully, without crashing. Readers who are experienced with making and flying radio-controlled airplanes will know that these are not toys, but rather a pretty tough technical assignment for beginners. Even if you buy a kit of parts and follow the instructions carefully, constructing them is difficult, as is starting a miniature two-stroke motor. Controlling the machine in the air is tricky too, until you get the hang of figuring out which way the plane is going and what the controls will do when you move them. It is unfortunate that flying model airplanes fell out of fashion in the 1980s, since they can give us insights into mechanics, electronics, and control, not to mention the physics of flight. It is difficult to think of any other "toy" that can provide such a satisfying technical challenge.

Luckily, the ekranoplan is not only simpler in design than a model airplane, but it is also easier to build, because it does not have to be built so light. You should try to keep the vehicle reasonably light in weight, of course. But you will

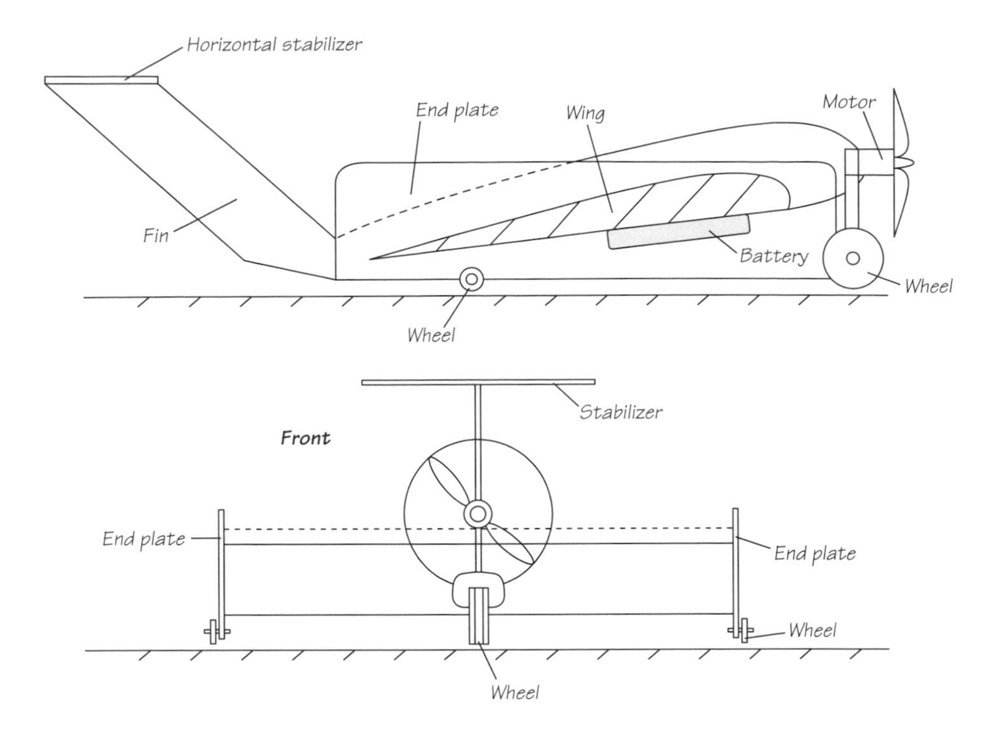

probably find that, at least on the smooth floor of a gymnasium, the vehicle will fly almost regardless of its weight. If it is heavier (within reason) it will simply fly lower. Nevertheless, I suggest that you use balsa and expanded polystyrene for as many parts as possible, using plywood fairly sparingly, and keep the weight down to around 1 to 1.5 kg (2 to 3 pounds) using the 540 motor suggested.

The simplest way to make the wing is with a homemade hot-wire cutter. Stretch a length of wire between two pieces of wood, spaced apart by the long strip as shown in the drawing, and keep it in tension with some twisted string. Then cut two pieces of plywood in the shape of an airfoil. I used the published shape of the Clark Y airfoil, which has the typical streamlined shape of a fish body on the top and a flat bottom. Gluing them together temporarily and then cutting them together will ensure that they are identical. Now separate them and glue them to each side of your chosen piece of polystyrene, so that they constitute "end-plate fins." (You will need to use a water-based contact glue compatible with expanded Styrofoam.) Then apply current from the 6-V supply to the NiCr wire and begin to the draw the hot wire over the two end cheeks. Twisting the string will increase the tension in the wire and make possible, within limits,

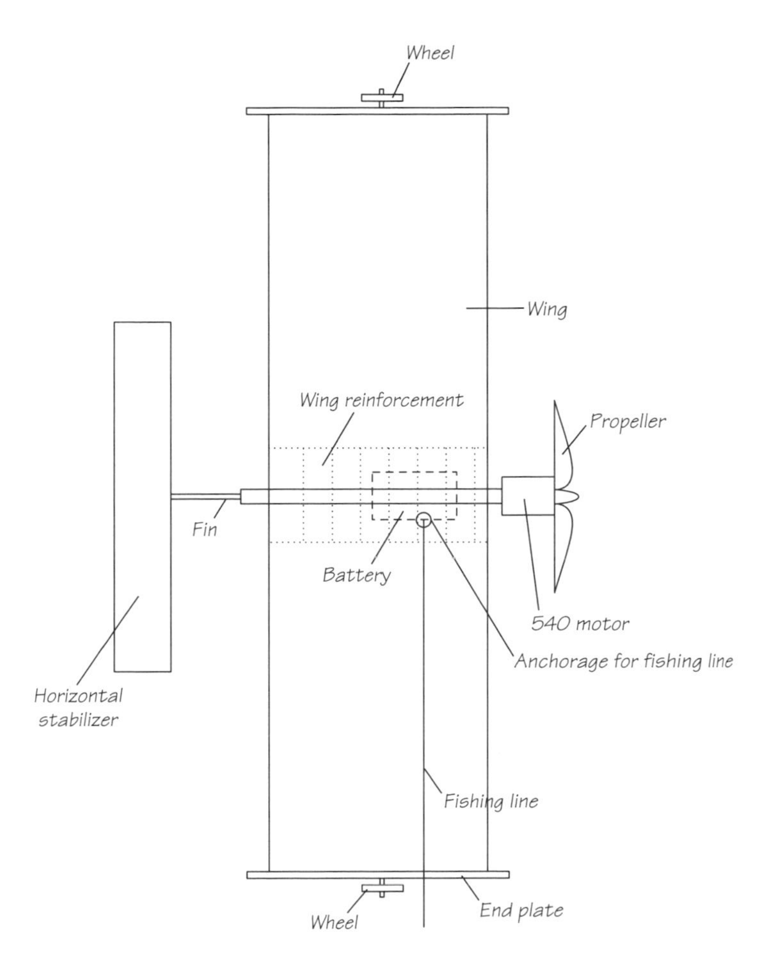

the reproduction of the end-plate profiles on the foam. Don't go too fast or you will distort the wire into an arc, and it won't faithfully reproduce the airfoil shape across the polystyrene.

Once you have cut the airfoil shape of the wing, remove the end-plate profiles and glue on the end-plate fins instead, using contact adhesive. The fins need to have a small wheel in the position shown, glued onto a small piece of plywood—a "hardpoint" or "anchorage" that will spread the load of the wheel axle over a larger area of the soft balsa. Contact glue the wing-reinforcing balsa sheet in the middle, top, and bottom. Now cut out the fuselage parts and glue them onto the wing reinforcement, top and bottom. You can now glue the motor to the front and glue the fin and stabilizer to the back. The single large wheel at the front, underneath the motor, needs mounting too. This wheel takes most of the machine's weight when it is rolling slowly.

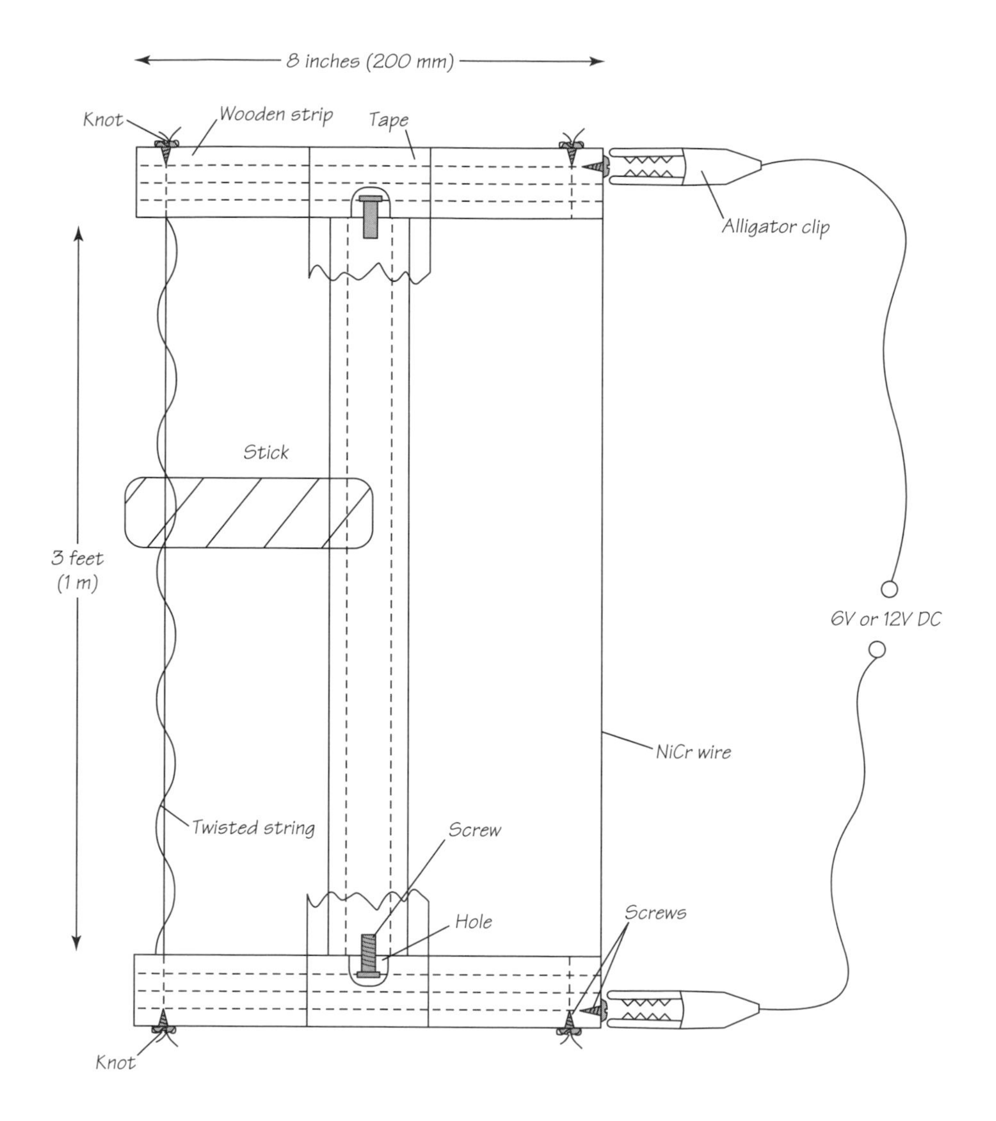

The horizontal stabilizer or tailplane has a strip at the back that is adjustable: I held it on by stiff wires, pushed and glued into the stabilizer and the strip. The fin has a rudder strip at the back, angled a few degrees from true to give the machine a tendency to turn away from the central post when it is operating. The battery is secured with elastic bands against an adjustment scale. The bands hook over wires that pass through the wing from small anchorages on the top. Finally, an anchorage point on the fuselage is needed to attach the tethering fishing line. The line passes through a reinforced hole in the wingtip. You should

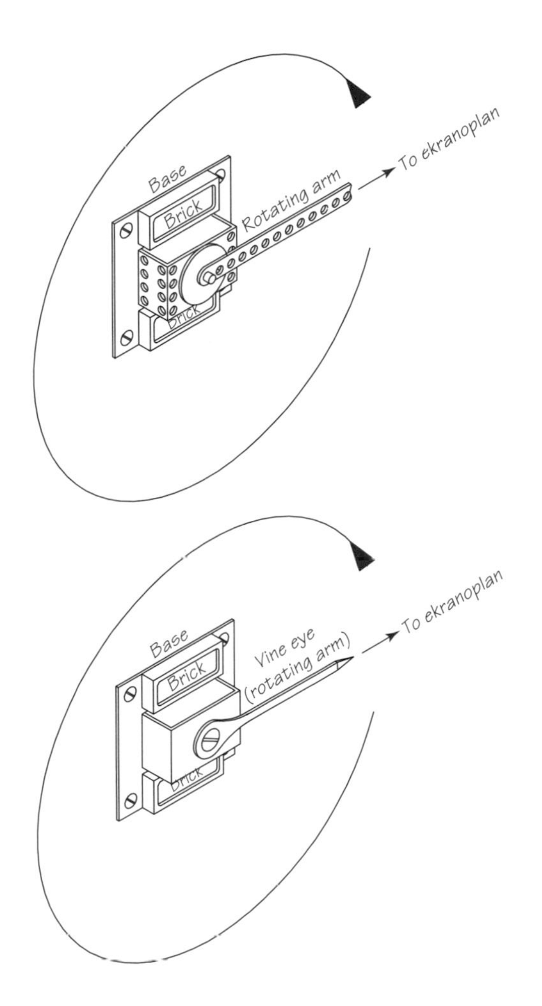

ensure, when fixing the position of this line, that the tether will not force the vehicle to fly at an angle to the ground: the best position is roughly at the middle of the wing chord, one-quarter of the way back from the front of the wing. The center of mass of the machine should also be adjusted until it is one-quarter of the way back from the front of the wing.

The central post, with its rotating arm on which to tie the tether, should ensure that the ekranoplan will fly in a polite circle, with the restraining line stretched out by centripetal force. You could use Erector set parts, or make the post from wood, with a projecting bolt on which the eyelet screw/vine eye will swing. Whichever you use, make sure that the arm turns freely on its bolt. Tie and glue one end of the fishing line to the eyelet screw. Now run out about 10

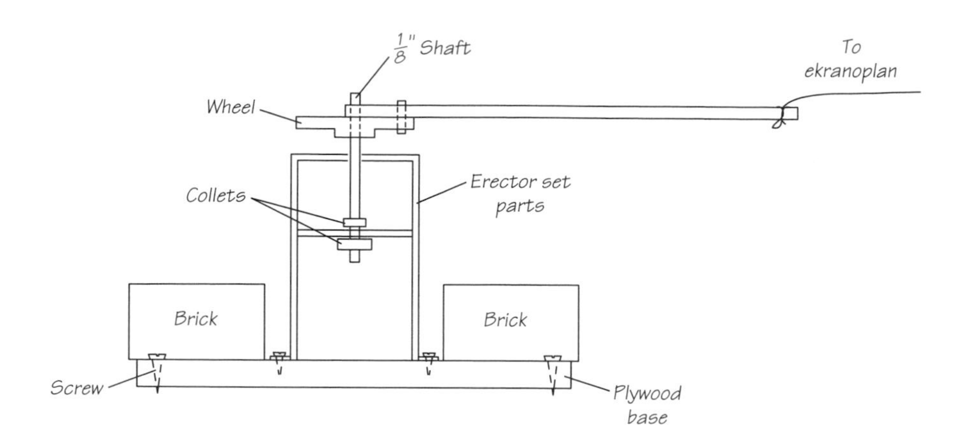

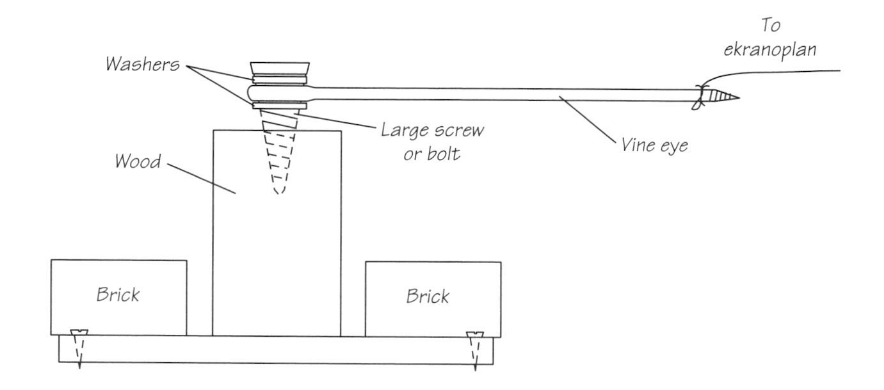

or 20 m (30 or 60 feet) of line to the ekranoplan and tie the tether line onto the craft. Walk the machine around in a circle to ensure that it will not collide with the walls or fence of your building or outdoor space.

What You Do

Try to find an indoor space in which to fly the machine. If you try it outdoors, then you will need to wait for flat, calm conditions: anything more than a slight wind will create problems.

The setting of the angle of attack (also sometimes called angle of incidence) on the wings is fairly critical. If it's too large, the vehicle may not accelerate to operational speed. If it's too small, the lift when the vehicle is rolling along will be insufficient for takeoff. Similarly, the position of the craft's center of gravity is critical. It is bound up with the relative angles of incidence of the wing and the horizontal stabilizer. Make all your adjustments little by little—otherwise you may skip past the optimum setting.

The two main means of adjustment for the ekranoplan are the adjustable elevator part of the horizontal stabilizer and the position of the battery. The former adjusts the angle at which the craft will fly—and hence the wing angle of attack—and is most effective at higher speeds. At lower speeds, the elevator will have less influence, while the position of the center of gravity (or center of mass), mostly determined by the battery position, will have more influence. This is why I suggest you make the position of the battery adjustable against a scale. Bending the elevator upward will, at high speeds, force the front of the ekranoplan up, and vice versa when you bend it down. Moving the battery backward will ensure that it will hold its nose up at low speeds.

Switch the power on, keeping your hands well clear from the high-speed propeller. Although only a few inches long, the propeller turns at an exceptionally high rpm and will give you a nasty cut if it catches your finger. This happened to me once, at a moment of maximum embarrassment: in front of an audience that I had just warned about the dangers of propellers! Worse still, I had decided to reduce the power available to the craft by putting the propeller on backward. This does *not* reverse the thrust from the propeller, which would only be reversed if the motor rotation was reversed. It does, however, make the propeller less efficient, because its airfoil cross section—normally blunt end upwind/sharp end downwind—is thereby reversed. This is what I wanted, and it did indeed reduce the thrust. It also meant, however, that the propeller, which would normally hit you blunt edge first, hit my finger sharp edge first, inflicting an unpleasant wound.

How It Works

After a slow run-up, the machine will lift off at low speed and then begin to speed up. If the tailplane is perfectly adjusted, then the speed-up process will continue, with the ekranoplan scudding along just 1 to 2 inches above the floor, increasing in speed to 25 or 30 miles per hour or more. You will find, by using different tailplane trim settings, that there are at least two other flight patterns. One has no takeoff at all: the machine whizzes around the circle rolling on its wheels, perhaps with the wheels only lightly touching the ground. The other has takeoff and acceleration, followed by a climb to an altitude of 6 m (2 feet) or more, then a slowing down (stalling) followed by a descent back to the floor, or maybe to within an inch or so of the floor. This pattern is followed by rapid acceleration at 1 to 2 inches' height until the speed builds and the climb begins again.

Whatever the flight pattern, you should notice that the ekranoplan flies fastest when it skims the floor, just an inch or less up in the air. It slows down if it flies lower and its wheels touch, so that friction slows it down, or if it climbs higher, when vortex drag slows it down. The acceleration of the craft when it gets close to the ground should be quite noticeable. You can see a similar effect working with another simple sheet-of-paper experiment. Drop a sheet of paper from about 60 cm (2 feet) onto a smooth floor. Depending upon how you drop it, the paper will likely glide one way a foot or two, then glide the other, rather slowly following a zigzag flight path until it gets to the floor. Then you will see the ground effect take control, and with luck your piece of paper will accelerate and zoom along a millimeter above the floor for up to 2 m (6 feet) or more.

If your ekranoplan flies lopsidedly, add small clay weights to the appropriate wingtip. I added a 13-g clay weight (about half an ounce) to the outboard wing, to compensate for the downward force of the string on the inboard wing. The horizontal stabilizer trim-strip adjustment mainly affects higher speeds; you can add small clay weights to the tail to get the craft to lift off at low speed, if that is a problem.

Measure how fast your ekranoplan can fly with a stopwatch, having measured the fishing line length. Once you have the machine working well, you can try adding further weights to the center of the ekranoplan to test its load-carrying abilities. You can expect it to fly lower if you add weight, but does it fly faster or slower?

THE SCIENCE AND THE MATH

The simplest way of considering ekranoplans or ram wings more generally is to think about them as dynamic versions of a hovercraft. The ekranoplan described here should have a surface area of around 0.3 m^2 and a mass of a little over 1 kg, including battery, trim weights, and so on. To keep it flying, the average pressure differential ΔP between the space underneath the craft and its upper surface needs to be

$\Delta P = Mg/A \sim 30$ Pa (0.3 mbar),

which is not a great deal of pressure. The upper surface of the wing will induce a lower pressure as it forces air traveling over the top as well, but let's assume for the moment that all the lift comes from higher pressure beneath the wing.

When an open-mouthed pipe faces directly into an airstream, air will tend to flow into it until, if the tube is blocked some way down, pressure builds up sufficiently to stop further inflow. This kind of arrangement, called a "pitot tube" or "pitot-static tube," is often used in aircraft to indicate airspeed: a pressure gauge is fitted to a tube connecting to the pitot and is calibrated to read directly in speed. Look on any aircraft and you will see a small pipe, just 2 or 3 mm ($^1/_8$ inch) in diameter and slightly tapered at the end, supported on a strut 100 mm (4 inches) or so high; that is the pitot tube. We can use the Bernoulli equation to calculate how much pressure builds. For example, a pitot tube facing into the air moving at 10 ms^{-1} would generate around 60 Pa of pressure:

$$\Delta P = \frac{1}{2}\rho v^2 \sim 60 \text{ Pa}$$

where ρ is the air density and v the airspeed.

Now picture a preposterously big pitot tube connected to a hovercraft's plenum chamber. Next imagine this whole curious assembly moving forward at zero altitude, and you can see that you have a strange sort of ground-effect aircraft. Air pressure accumulated by the forward motion of the pitot is communicated to the plenum and lifts the whole craft up from the ground. The ekranoplan differs in practice from this conceptual model, but not in principle: its air intake is a low rectangle rather than a circular shape, facing forward as required. And it has serious air leaks around its plenum—underneath the end-plate fins and at the trailing edge. But ignoring these differences, we should expect the ekranoplan to take off at a relatively low speed, maybe as low as 7 ms^{-1}, based on the above calculations. We can combine the two equations for a simple ekranoplan equation:

$$v_{\text{l-o}} = \sqrt{(2Mg/A\rho)}.$$

A lower liftoff speed $v_{\text{l-o}}$ can be achieved by increasing the wing area or decreasing the craft's mass. The formula says nothing about the details of the ekranoplan design such as the wingspan, chord, or angle of incidence. It does assume that the leaks from the edges of the craft can be made smaller than the inflow, so that the pressure available is not decreased too much by leakage, and these considerations do enter into the aircraft design. Our design has a substantial frontal area to capture a large flow of air as it moves forward, and it at least attempts to seal off the leaks by having end-plate fins and sweeping the trailing edge low to the ground.

Another way to think about the ekranoplan is as a low-flying airplane. An ekranoplan produces less drag than a regular wing in free air for the same lift, because it does not generate the large wingtip vortexes that a normal wing does. Most of the energy wasted by an airplane because of drag goes into the formation of two very large vortexes, one from each wingtip.

Why should a vortex form at the wingtip? Because air pressure is higher beneath the wing than it is above. Air will try to flow from bottom to top; along the length of the wing this is prevented by the speed of flow around the trailing and leading edges, and by the wing's simply being in the way otherwise. At the wingtip this scenario breaks down because flow can take place around the tip. Combine the circulating motion from below to above the airfoil at the wingtip with the forward motion of the aircraft, and you have a helical motion: a vortex. You can see these vortexes forming at the wingtips of commercial jets when they land and take off in damp air. Look through a pair of binoculars, and you will see that the contrails left by jets flying at high altitude are also tip vortexes.

The ekranoplan wing does not form a tip vortex, because the ground below prevents its formation: the air cannot flow from one side of the wing to the other behind the craft freely. This is the key to its advantages over a normal airplane. The lift force—the useful effect of a wing—can be expressed approximately by

$$L = \frac{1}{2}\rho v^2 A C_l,$$

where A is the wing area and C_l is the lift coefficient. C_l depends upon the angle of incidence of the wing and tends to a broad maximum at around ten degrees. Meanwhile the drag force—the force required to move the wing forward (the problem effect of a wing)—can be expressed by

$$D = \frac{1}{2}\rho v^2 A C_d,$$

where C_d is the drag coefficient. The value of C_d in certain ranges of speed depends largely upon the formation of the tip vortexes. C_d varies with the angle of incidence but is often fairly constant until about 10 degrees, when it rapidly increases, creating the condition known as "stall," where the drag arises from the formation of a set of turbulent vortexes smaller and much more numerous than the tip vortex. Lift and drag coefficients are approximately constant for a certain angle of incidence and certain ranges of wing size and airspeed.

The "figure of merit" often chosen for an airplane wing is its lift/drag ratio. The L/D is simply given by $L/D = C_l/C_d$. With a conventional airfoil section, it often reaches its maximum at an angle of incidence of around 10 degrees, since beyond this angle the drag increases rapidly and lift no longer increases much, if at all. The drag force coefficient C_{di}—often called "induced drag"—that results from vortex formation can be approximately given by

$$C_{di} = kC_l^2/\pi R,$$

where k is a constant and R is the aspect ratio (wingspan/wing chord). Thus, if all the drag force on an airplane were given by the vortex effects, we would have:

$$L/D \sim \pi R/kC_l,$$

a ratio still dependent on the lift coefficient, which as explained varies with angle of incidence. It shows, however, that for a normal airplane wing, changing to a high aspect ratio yields a big payback because it greatly reduces drag, other things being equal. For our ekranoplan, however, the vortex drag can be neglected, because the end-plate fins almost completely stop the formation of a vortex at low flying heights. The ekranoplan has the equivalent of a wing with an infinite aspect ratio from the point of view of vortex drag. This, then, is the source of its efficiency: it should be able to travel much faster than the normal finite wingspan. This is also what is meant by the "ground effect": the acceleration you see when an ekranoplan dips closer to the ground can be thought of as stemming from these dynamics.

One danger that a ram wing craft may be subject to is that of Venturi (or Bernoulli effect) suction. If the vehicle gets too close to the ground, the air flowing swiftly beneath it will tend, despite the presence of the rapidly "moving" ground, to act as constriction, accelerating the air. That acceleration can only occur if there is a low pressure, and that low pressure will mean that the vehicle is sucked down. The suggested angle at which the wing is set in our ekranoplan, along with its spacing from the ground due to the depth of the end-plate fins, will probably ensure that this does not happen.

If you want to try some more ekranoplan calculations, you may come across the difficulty faced by many engineers and scientists working with relatively small aerodynamic systems: at normal speeds, the Reynolds number Re is low:

$$\mathrm{Re} = \rho L v/\mu,$$

where L is the characteristic length scale and μ the air viscosity. The Re gives the ratio between inertial forces divided by viscous forces: at low Re, flows are smooth, like those of molasses. At high Re, flows are turbulent, easily forming eddies and vortexes in any size. Whereas a jet plane might have Re = 500,000,000, a flying duck might have Re = 300,000, and our ekranoplan would yield a similarly low value. Aerodynamic systems with a similar Re behave in similar ways with respect to flow patterns. This is the reason that small wind-tunnel models of aircraft can be made to simulate full-scale behavior. But the smaller size of L means that the tunnel must be made faster (v larger) or the air density ρ higher (e.g., by using cold nitrogen gas generated by boiling liquid nitrogen). A low Re means that you cannot use the data gathered in textbooks on full-size aircraft; you have to look for special low-Re data. Authors such as Martin Simons (see the references at the end of this experiment) have investigated model planes specifically, while Steven Vogel describes nature's flyers and well demonstrates the effects of low Re on aerodynamics.

And Finally . . . Flying Skirts

At first glance, you might think that any ram wing inevitably needs to have a wing of some kind and a stabilizer of some kind, and would thus always look something like our ekranoplan . . . but does it? Maybe you could build an even less conventional device. Could you, for instance, build a "train" of ram wings, with the leading powered ram wing pulling a lot of unpowered ram wings behind? Or could you build an ekranoplan that used a water screw for propulsion? A water-powered ekranoplan might have a degree of automatic regulation of its flying height: if it flew too high, the water screw would pull out of the water, the propulsive force would stop, and the craft should then glide down to the correct operational height. But would such an automatic regulation actually work in practice? Would the propeller work if only a single blade projected into the water at a time? This is how some high-speed hydroplanes work: they use a two-blade propeller mounted so that at speed only one blade will reach the water at a time, an arrangement sometimes described as a "single-blade propeller."

You could change the proportions of the ekranoplan's wing considerably, and it would still probably work well. With a long span and a small chord wing, the wingtip vortexes should be less significant, reducing the size of the ground-effect boost: in the limit, you would have a very low-flying aircraft that lacked only the controls and the power to go higher. With a shorter span and a longer chord wing, which could have a chord longer than its span, the opposite applies. You would have a machine that was much more confined to operation quite near the ground.

That being the case, could a low-speed ekranoplan use a flexible skirt, like a hovercraft? The skirt would have to be open at the front, of course, to admit air. And the side skirts would need restraints to ensure that they were not pushed out by the internal pressure at speed. The skirts would have to be made of light, thin, highly flexible material; very thin polyethylene (PE) film, the sort used for covering furniture when you are painting a room, might do. And we would need a new name for the craft. Anglo-Saxon enthusiasts might prefer the prosaic "skirt-

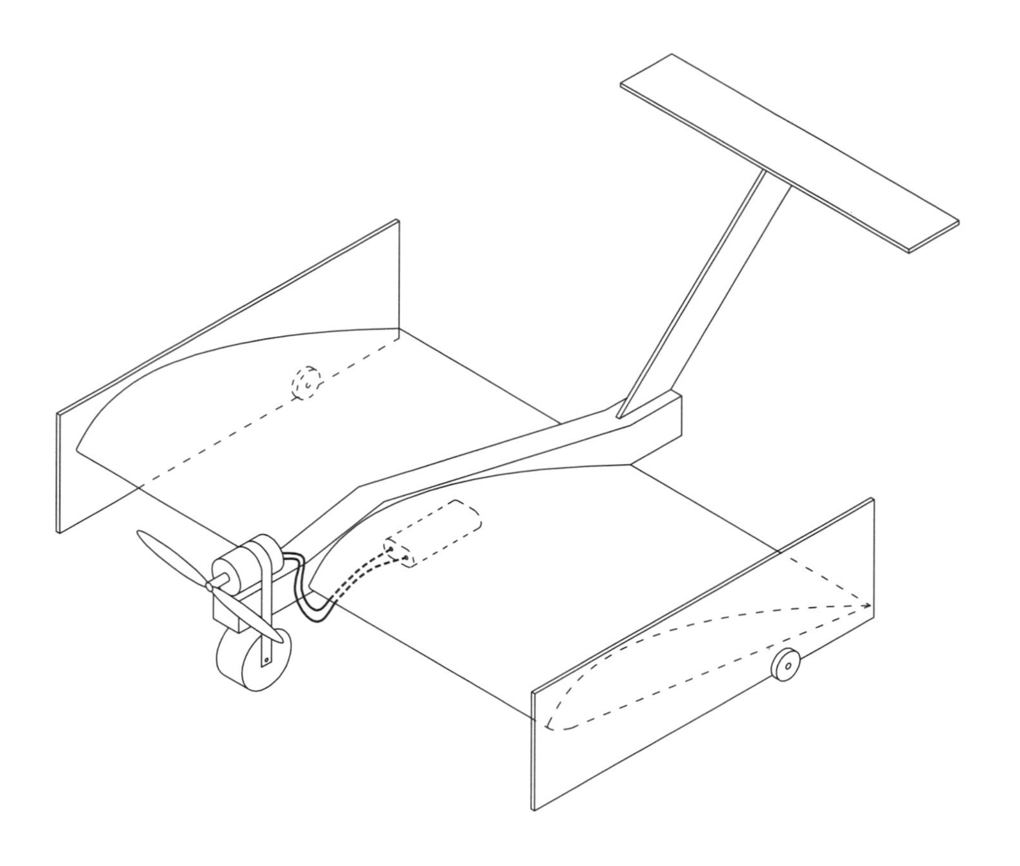

wing," but this sounds too much like an insect, the type you wouldn't want around your barbecue. How about "pterofousta" (Greek for wing-skirt), "jupewing" (from the French for skirt), "kiltoplane" (from the Gaelic) or even "faldavolante" (from the Spanish for flying skirt), with overtones of that famous image of Marilyn Monroe?

REFERENCES

Simons, Martin. *Model Aircraft Aerodynamics.* 2nd ed. Hemel Hempstead, Hertfordshire, U.K.: Argus Books, 1987.

Vogel, Steven. *Life in Moving Fluids: The Physical Biology of Flow.* 2nd ed. Princeton, N.J.: Princeton University Press, 1994.

11 Slimemobile

> We are far more concerned with research and the behavior and design of these craft at these speeds than we are with merely breaking the world record.
>
> —Donald Campbell, who later garnered the world water speed record of 202 mph at Lake Ullswater, U.K., on July 25, 1955

Rest assured that we won't be breaking any speed records with the slimemobile, so we will indeed concern ourselves with its behavior and design, based on the fundamentals of lubrication science, or tribology. Two dry surfaces rubbing on each other exert a friction force. This force of friction still contains some mysteries today, hundreds of years after scientists (including the illustrious Isaac Newton) first investigated it. But we can certainly say that some of the force of friction arises from the work that must be done to unweld minuscule pressure points on the two surfaces that have welded. Another part of frictional force derives from the work done in jigging the surfaces as they ride up and down tiny ridges and troughs on each other's surfaces. The net result is the production of heat where the surfaces rub, which wastes large amounts of power.

With two fully lubricated surfaces—surfaces between which there is an unbroken layer of lubricating fluid—the results are quite different. Fully lubricated surfaces slide smoothly and freely, with only minute amounts of heat generated; such heat is carried off by the lubricating fluid, which thus acts as a coolant.

In this project we create, in effect, a planar version of a hydrostatic bearing. Most oiled bearings or sliding surfaces are only dynamically lubricated. Stop all the movement, and the oil between the surfaces will squeeze out and the bearing or slider will be stuck, almost as if the oil had not been there at all. It is only once the bearing surfaces are moving at a reasonable speed—dragging oil into the interface—that the bearings show the low friction of which they are capable, thus revealing the oil's true benefits. In a hydrostatic bearing, by contrast, we actively pump in the lubricant to the interface, so that, even with the surfaces stationary with respect to each other, there is only the viscosity of the lubricant to restrain motion.

What You Need

- ❑ Large, shallow jar lid
- ❑ Large syringe (20 ml or more)
- ❑ Strong rubber bands
- ❑ Motor
- ❑ Battery/battery box
- ❑ Propeller
- ❑ Short piece of flexible plastic tubing to connect syringe with plenum
- ❑ Jell-O or similar gelatin
- ❑ Level glass surface measuring 1–2 × 1 m (3–6 × 3 feet)
- ❑ Fence wire
- ❑ Wood
- ❑ Glue

What You Do

The slimemobile is a kind of hovercraft whose propulsion function is handled in the normal way by an airscrew or propeller, but whose hover function is provided by injecting a high-viscosity liquid slime. The slime is injected into the plenum beneath by a spring-loaded syringe.

The plenum of the slimemobile is a large, shallow lid from a jar or paint can. Make sure it has not been twisted. Drill it in the middle and fit a short piece of plastic tubing, the right size for the syringe to be a good leak-tight push fit, fixing it in place with hot-melt adhesive. Glue the motor/propeller onto a wood pylon, as illustrated, above a rechargeable battery supply. The wire tail—

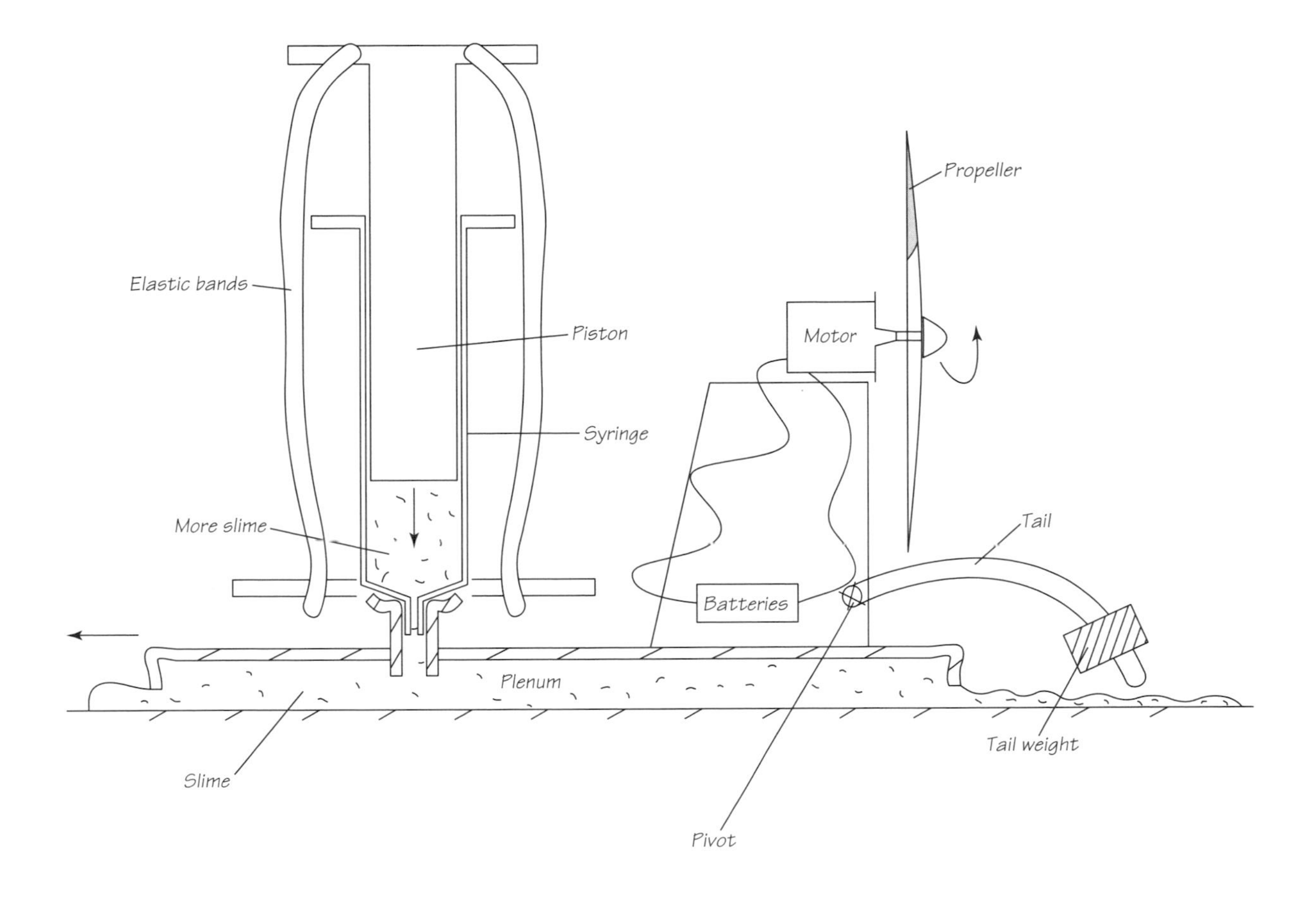

a weighted piece of wire, pivoted on the back of the pylon—drags behind the device so that when the propeller blows, it will push the slimemobile forward. Move the propeller as far forward as possible, while maintaining the balance of the slimemobile. This will make the path it follows quite straight; you may even find the tail unnecessary.

When assembling your vehicle, make sure that the components are reasonably well balanced. Support the jar lid with a pencil in "teeter-totter" mode, checking that you have achieved balance in both fore-and-aft and lateral directions.

All this is pretty straightforward, but the slimemobile won't work without slime, and it will move only as long as its slime supply lasts. So your first priority is to make the necessary slime, and then experiment to generate a steady supply of slime from the syringe for the necessary amount of time. I found that I could get about a minute's worth of slime by using a standard 20-ml syringe propelled by elastic bands, the syringe filled with Jell-O made with water to more or less standard proportions or with a little excess water, kept at room temperature so that it will only just set.

With the vehicle complete, you will need a flat glass surface on which to try it. It needs to be absolutely flat, so check the slope with a spirit level. Now switch on the slimemobile and fill the syringe with slime. Push the syringe tip into the pipe on the plenum, place the slimemobile on the glass surface, and pull the elastic bands over the top of the plunger. With luck, the plenum will quickly fill with slime. The slimemobile will then lift up a fraction of a millimeter, and the propeller will push it along.

The slimemobile stops quickly once the syringe reaches its end stop. Without the lubricant pumped in under pressure, the plenum sits down, and static solid-on-solid friction—or something quite close—takes over. The feeble thrust from the small fan is no longer capable of propelling the slimemobile forward.

THE SCIENCE AND THE MATH

Even if you cover the entire sheet of glass with slime, the friction between the vehicle and the glass is large enough that the slimemobile will not move unless the slime pressure exceeds the weight/area of the vehicle. At that point the slimemobile "lifts off." However, as soon as the slime lifts the "skirts" of the jar lid off the glass substrate, you have a whole new ball game: now the only force restraining the slimemobile's forward motion is the viscosity of the slime underneath the skirts and the rest of the plenum.

There is a fundamental limit to how far a slimemobile can travel, because it will run out of slime. With no slime supply, the slimemobile will not move in the slightest: the feeble force applied by the low-power propeller will be far too weak to shift it. There is another fundamental limit on how fast it can go, because the speed is restricted by the slime-feed rate. In fact, if these two variables are proportionally related, then the distance the slimemobile can travel would be simply fixed by its onboard slime supply. But is that how the math works out?

Most of the viscous forces occur where there is a large velocity gradient—a large differential in velocity across a thin layer. So let us first make the assumption that most of the drag occurs within a short distance of the lip from the surface across which it is creeping. The slime has a very high viscosity, but only at the lip, where the distance is small, does it get a chance to slow down the slimemobile, as we can see by looking at the ordinary viscous equation

$$F_{drag} = \mu A \, dv/dx,$$

where F_{drag} is the drag force, μ is the viscosity of the slime, A is the area over which the slime layer acts, and dv/dx is the velocity gradient, which has an average value equal to the slimemobile's speed divided by the lip clearance from the ground, h. The area A is simply given by the perimeter of the slimemobile, whose base diameter is R, multiplied by the width of the lip L, so $A = 2\pi RL$. So we now have

$$F_{drag} = 2\mu\pi RLv/h.$$

As our propeller-motor-battery combination gives a thrust force F, the maximum speed v_{max} of the slimemobile (once it has accelerated to full speed!) will be given by turning the above equation around, in other words:

$$v_{max} = Fh/(2\mu\pi RL).$$

The slimemobile goes faster if we add more thrust, and slower if we make it larger in radius or make the lip width L larger. Now all these quantities are known except h. We should expect h to vary with some of the other terms in the equation above. For example, if we make the slime viscosity smaller, then the drag will decrease, but so will the height h, since the slime will run out of the gap quicker. So we need to calculate the "hover altitude" of the slimemobile.

We can use another viscous flow calculation to estimate h as follows:

$$Q = (1/12\mu L)wPh^3.$$

This gives the volume flow rate Q of slime from a slot with length in the flow direction L, with a width w much greater than its height h, under pressure P. In our case, the slot's width = $2\pi R$, so we have

$$Q = (^1/_{16}\mu L)\pi RPh^3,$$

which can be turned around to give the hover altitude h:

$$h = (6\mu LQ/[\pi RP])^{1/3}.$$

So we can now re-express v_{max}:

$$v_{max} = F(6\mu LQ)^{1/3}/([\pi RP]^{1/3}[2\mu\pi RL])$$

which gives us

$$v_{max} = F(0.75Q)^{1/3}/(\pi^{4/3}R^{4/3}\mu^{2/3}L^{2/3}P^{1/3}).$$

We know that the pressure P that the slime exerts over the whole area of the base must be equal to the slimemobile weight Mg, in other words,

$$P = \text{Weight/Area} = Mg/\pi R^2$$

which finally gives us (phew!) . . . the slimemobile equation:

$$v_{max} = F(0.75Q)^{1/3}/(\pi R^{2/3}\mu^{2/3}L^{2/3}M^{1/3}g^{1/3}).$$

Does this equation seem reasonable? If you increase the propeller thrust F, then you go faster; likewise if you increase the slime flow Q, although your speed will increase only slightly. In the opposite direction, increasing the radius R of the slimemobile base will decrease the speed, although only with a two-thirds power, the same power applying to increases in viscosity μ and to increases in the lip width L. Increasing the mass M of the slimemobile will actually decrease the maximum speed only a little, at one-third power.

Of course, we can use dimensional analysis to check that the formula is reasonable (see the Hints and Tips section at the back of the book):

$$v_{max} = F(0.75Q)^{1/3}/(\pi R^{2/3}\mu^{2/3}L^{2/3}M^{1/3}g^{1/3}).$$

		0.33	**0.33**	**−0.67**	**−0.67**	**−0.67**	**−0.33**	**−0.33**	
		MLT−2	**L3T−1**	**L**	**ML−1T−1**	**L**	**M**	**MLT−2**	
v	=	F	Q	R	Mu	L	M	g	
M		M	M	M	M	M	M	M	**Total dimension**
0.00	=	1.00	0.00	0.00	−0.67	0.00	−0.33	0.00	0.00
L		L	L	L	L	L	L	L	L
1.00	=	1.00	1.00	−0.67	0.67	−0.67	0.00	−0.33	1.00
T		T	T	T	T	T	T	T	T
−1.00	=	−2.00	−0.33	0.00	0.67	0.00	0.00	0.67	−1.00

Naturally there is more depth to the analysis we could achieve with the slimemobile. The flow through the syringe nozzle will clearly vary as the tension provided by the rubber bands decreases. Since a lower flow will equate to a lower "hovering height," and lower hovering means higher drag, we should expect the slimemobile to slow down with time.

And Finally . . . Slime Races

What about nature's slugs and snails (generically called mollusks): how do they move, and what does their slime do for them? They are fascinating creatures, once you have gotten over the "ugh" factor. Some mollusks can live for a long time—100 years have been recorded—and are hermaphroditic (both male and female). I have found *Helix pomatia*—larger and more striped than the normal English snail—in the hills near my house. According to legend, these edible snails were brought over by the ancient Romans, who considered them a delicacy, cultivating them in little gardens called *cochlearia*. They are thus to this day found mostly near old Roman roads and towns.

According to Vines and Rees (see the reference below), molluscan slime is a genuine lubricant. The motion of these animals involves a wavelike movement of what amount to exceedingly short legs. Muscles in the foot cause ridges to be formed that ripple along the ground, behaving rather like the waves of leg movement in a centipede, except that the "legs" are about 10 mm wide and 0.5 mm long, with "feet" only 0.5 mm long. These tiny leg ridges give snails and slugs a maximum speed of only 10 cm per minute or so, which certainly makes our slimemobile a potential world-record machine!

How about a slimemobile race? How could you best give your slimemobile the edge in performance against others of the species? More motor power sounds promising—but that might mean a heavier motor. More slime flow might give you a larger hover height and lower drag, but more slime means a larger weight of slime to carry, which would lower hover height and may mean a higher slime flow is necessary.

REFERENCE

Vines, A. E., and N. Rees. *Plant and Animal Biology.* 4th ed. London: Pitman, 1972. See pp. 1286–87 on locomotion in slugs, snails, and similar creatures.

12 Ball River Bobsled

I realized that . . . the real barrier was not in the sky but in our knowledge.

—Chuck Yeager, pioneer aviator who broke the sound barrier and piloted his rocket to supersonic speeds

"Oiling the wheels of industry" is a familiar turn of phrase. In fact, however, the wheels of industry today rely more on roller and ball bearings than on lubricating oil. Wheels on cars and trucks turn efficiently because of the low friction that rolling bearings allow, and rolling bearings will in many cases run quite well without oil. Roller and ball bearings work by swapping the typically high forces of sliding friction for the low drag forces of rolling friction. With very hard balls or rollers (sometimes even made from ceramics) and surfaces, this type of bearing can be extraordinarily efficient. These facts were known a long time ago. Leonardo da Vinci, for example, incorporated them in some of his designs. He proposed the use of roller bearings for the suspension of a giant bell weighing many tons. The idea was that the bell, despite its Brobdingnagian proportions, could still be rung easily by a single person. For long centuries, however, the materials needed for effective ball bearings were simply not available. Hardened steel and precision machining were needed before the idea could leave the drawing board and be used in real machines.

Although 99 percent of roller bearings are used in rotational motion, the same principle can be applied to linear movement. There are various designs involving both rollers and balls that allow smooth, almost frictionless linear movement. If you are sitting at a desk with a large, heavy drawer for files, pull the drawer out and examine it carefully: many office desks include linear roller bearings for their main drawers. Another place you may see such linear slides is in rowboats, particularly the lightweight racing rowboats such as the sculls or "racing eights" used on rivers. Several decades ago, rowers would have pulled at the oars with their hands alone. Today, to extend and to add power to the strokes, they can shift their seats and use their thigh muscles to pull the oars. The sliding seat must be as nearly frictionless as possible, so as not to waste the rowers' precious muscle power, and this result is achieved by using roller slides.

What You Need

- ❏ Plastic U-channel, 85 or 100 mm (3½ or 4 inches) wide and 4 m (12 feet) long, or as long as you can afford or fit in your room or backyard
- ❏ Small (3-mm) spherical glass beads
- ❏ Drainpipe 75, 100, or 150 mm (3, 4, or 6 inches) in diameter, to be sawn as described below, radius of curvature to be less than that of U-channel
- ❏ Metal saw

Jet Propulsion

- ❏ Short section of pipe, 12 mm in outside diameter
- ❏ Wooden block
- ❏ Balloons
- ❏ Plug

Propeller Propulsion

- ❏ Small propeller, or larger propeller cut down
- ❏ Motor
- ❏ Battery
- ❏ Battery holder

What You Do

The ball river bobsled is driven forward, like the slimemobile, by an airscrew or propeller pushing air backwards, or by a jet of air from a balloon. However, the friction between the bobsled and Mother Earth is lowered, not like a normal bobsled, by ice, but by tiny glass balls. These are sprinkled on a U-channel that retains them, and they roll a little as the bobsled passes over them.

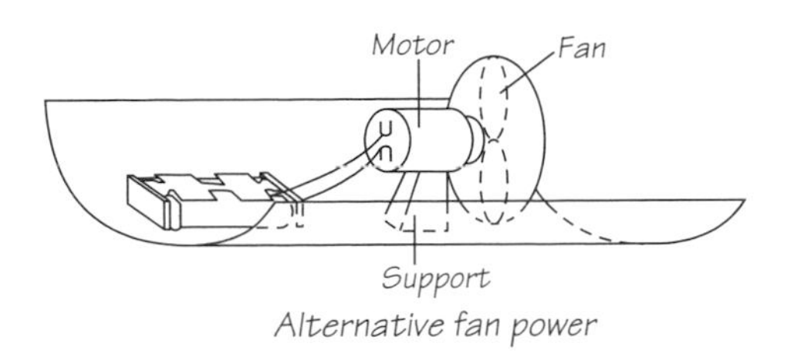

Alternative fan power

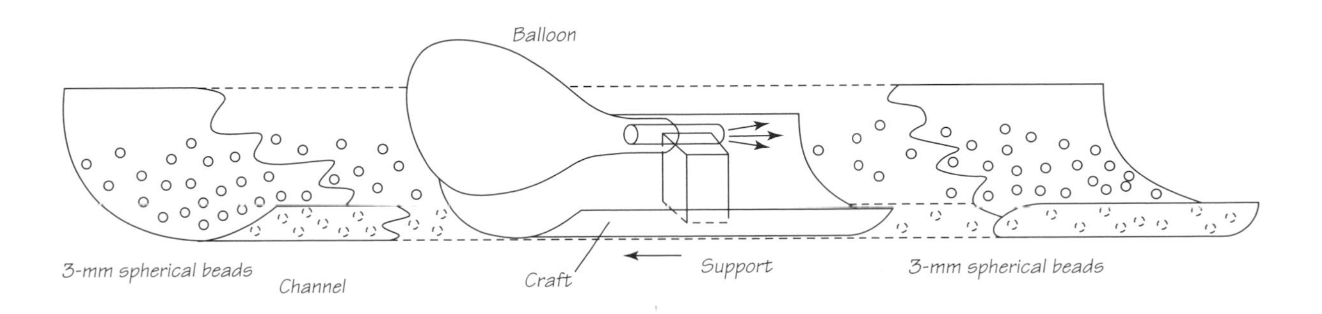

The shallow U-section that forms the base of the bobsled must be of a slightly smaller curvature than the U-section track. I could not find a smaller U-channel that would fit inside the track U-channel, but I did find a section of pipe that was about right. So if you can't find a ready-made U-channel, buy a drainpipe and saw a section of it along its length, so that you'll have a piece whose cross section is about one-third of a circle. As suggested in the parts list, you could choose to make either a simple balloon-jet drive, or a rather more predictable continuous drive using a small propeller and electric motor. The diagram shows the kind of arrangement you could use, although no doubt other configurations would work too.

Some key features of your ball river bobsled: first, the vehicle's center of gravity must be lower than the former axis of the pipe section. Otherwise, it may capsize. Second, its radius of curvature must be comfortably smaller than that of the drain channel in which it will run. Given the thickness of the drain channel itself (1.5 mm) and the size of the balls (3 mm), the bobsled's radius should measure at least 6 mm less than the channel's. The balls I used are sold by chemical manufacturers as inert filler material in chemical reactor beds and the like. Because they are used in huge numbers, they are made by efficient mass-production techniques and are quite inexpensive. But you could use any small, hard spheres—even ball bearings—if you can lay your hands on enough of them.

When all is ready, load your drain channel with the glass balls. Now start the motor or inflate the balloon, and set the ball river bobsled going.

How It Works

The balls themselves do not, unless pushed along in the bow wave, travel much along the channel. Provided that the channel is level and that traffic passes evenly in each direction, they stay largely where you put them, moving along by at most half of the vehicle length in the direction of travel, and the same distance back again with the reverse run that follows. They thus exhibit a kind of "tidal flow" with each passage to and fro of the bobsled.

The vehicle has very low effective friction and should run reasonably fast, but it has a quite well-defined maximum speed and does not get any faster, even on a long section of straight track. It does not, as you might expect for such a low-friction device driven by a high-speed jet or propeller, accelerate continuously to higher speeds. Instead, the bobsled loses energy to the motion of the glass beads, and this is what limits its speed. We explore this phenomenon in more detail below.

Try larger and smaller numbers of balls. With small numbers, the vehicle will run roughly; with too many, a "bow wave" may build up. How critical is the size of the beads? Will smaller beads allow faster speeds, or would larger beads be better? Clearly, if the channel is slightly rough, then larger beads will run more smoothly over the roughness. But if the drain channel is truly smooth, won't smaller beads work better?

THE SCIENCE AND THE MATH

As the vehicle slides by at full speed, the channel in which it runs does not move at all, while the glass balls move at half speed: they rotate such that their upper contact points move along at full speed, while their lower contact points stay at rest, and thus their centers move at half speed. The same geometry applies to other linear slides such as roller-drawer slides or the sliding seat in a "racing eight" boat. Look at your desk, or at a sculling boat, and you will see that the drawer in the former and the rower's seat in the latter move a certain distance, but the rollers themselves move only half as far.

What about the inertia of the ball river bobsled? Clearly, the inertia of the vehicle body itself can be simply expressed by the vehicle's mass. However, there is an additional amount of inertia owing to the fact that when you accelerate the vehicle to velocity v, you also accelerate the set of balls underneath it to velocity $v/2$.

We can calculate the effect of this dynamic as follows: Suppose that there are approximately N balls of mass M_b per unit area, and that the bobsled area is A. The effect of the balls is equivalent to adding a mass of half the mass NAM_b of the NA accelerated balls to the vehicle mass M. The bobsled thus behaves like a vehicle of mass M_{eff}, where

$$M_{eff} = M + \tfrac{1}{2}NAM_b.$$

The ball river bobsled has just a little more inertia and is therefore a bit more difficult to accelerate than a bobsled on ice would be. This is not the whole story. There is also a constant-energy-loss mechanism, analogous to friction. The glass balls set into motion by the bobsled still move afterward. They do not pass their energy to the glass balls in front, or at least not with any efficiency. There is, in effect, an additional drag force on the bobsled. (Clearly, the "bow wave" effect we see when the bob runs is passing some energy to the balls in front: equally clearly, however, the balls behind are still jiggling and left with some of the energy that went into moving them.) There is a constant loss of energy, even when the bob is running at a constant speed. We can estimate this loss by making the simple assumption that the balls are at zero speed before the front of the vehicle touches them, and that they have the kinetic energy due to motion at speed $v/2$ once the vehicle has passed:

$$KE \text{ of each ball} = \tfrac{1}{2}M_b(v/2)^2 = \tfrac{1}{8}M_b v^2.$$

At speed v, if the bobsled has length L, it will create new balls at this KE at a rate of NAv/L per second, assuming that NA balls are supporting it at any one time. So KE lost per second = "effective frictional power" = $\tfrac{1}{8}M_b NAv^3/L$. This amounts to a drag force that is proportional to v^2, similar to the aerodynamic drag on an object at a high Reynolds number, such as we can observe with airplanes and automobiles. As with aerodynamic drag, this force is small at low speeds but becomes dominant at higher speeds.

The bobsled also sets the balls into rotation. You can show, however, that this does not have a large effect on the bobsled's acceleration. You might comment that the rotational inertia of the wheels is also a factor in ordinary wheeled vehicles, and you would be right. This is one good reason for making any vehicle's wheels as light as is practical (the other reasons have to do with cornering: heavy wheels will momentarily bounce and lose their grip on corners). Look inside a car showroom, and you will see lots of vehicles with lightweight magnesium alloy wheels. However, the rotational inertia effects on acceleration are relatively small in magnitude both for ordinary vehicles and for the ball river bobsled.

In the case of the bobsled, rotational inertia affects not just acceleration, but also the energy losses from running at a constant speed, and these losses are substantially affected. You can show that the effect of the energy stored in rotation makes the balls behave as if they had a larger effective mass M_{beff}:

At velocity v, KE of the center of mass of a ball $= \tfrac{1}{2}M_b v^2$.

At velocity v, the rotational speed of the ball $\omega = v/R$, and moment of inertia $I = \sum (mR^2) = \tfrac{2}{5}M\,R^2$.

Hence, at velocity v, rotational KE of the ball $= \tfrac{1}{2}I\omega^2 = \tfrac{2}{10}Mv^2$.

This means that $M_{beff} = \tfrac{7}{5}M_b$.

It is this M_{beff} that really should be used to estimate the drag force effects. So

$$\text{Drag power loss} = \tfrac{7}{40}M_b NAv^3/L.$$

And Finally . . . a Ball River Barge

What about cornering? Home Depot and similar stores offer a limited selection of parts suitable for simulating a bend in your river; you might experiment with some sharp angles and knees designed for roof gutter channels. You can often buy more gently curved pieces approximating 45-degree or 30-degree angles at specialist vendors. But can your ball river bobsled really go politely around corners at speed, or will it drive off the balls in the middle of the channel, following centripetal forces, and run aground without the balls beneath it?

Could the bobsled effect be used for high speeds at a larger scale? Indeed, could a high-speed train of the future use these principles? What about a heavy-weight ball river bobsled? The system would appear to lend itself well to very heavy loads moving at low speed. But how do you get the pulling power to tug a heavy bobsled along? Could you tow it with a horse, like the river barges of history? Perhaps a more intelligent approach is possible: an equivalent to the river-barge water-screw propeller. Could a kind of paddlewheel, for example, be used, batting ball bearings backward to urge the bobsled forward? Wouldn't the necessary backward velocity have to be enormous to generate enough thrust? Maybe a paddlewheel with points on it would be a better idea. This could push aside ball bearings to contact the channel directly and thus generate a forward thrust, in a manner something like poling a punt along a river.

13 Skidmobile

> I've always had an idea that my retirement would be the greatest contribution to science the world has ever known.
>
> —Groucho Marx, *Animal Crackers*

It is a curious fact that 99.999 percent of the time, road vehicles run straight along the street without any hint of skidding—except in the movies. Actors, it seems, are the world's worst drivers, effortlessly defeating the antiskid devices developed at great expense by the auto companies. Once on a movie set, vehicles spend half their time sliding at peculiar angles, with their wheels spinning and the burned rubber emitting clouds of smoke.

Here we build a machine that arguably represents the logical conclusion of the history of movie vehicle development: an automobile that skids every moment of every journey. The Marx Brothers once appeared in a film with an extraordinary car designed to be driven in a peculiarly democratic way. It had four wheels, one on each corner like a regular car, but the difference came in the control department—it had four steering wheels as well! I think that the skid-mobile is a vehicle worthy of the Marx Brothers.

What You Need

- ❑ 2 or 3 motor-transmission units
- ❑ 3 smooth wheels
- ❑ Model servo unit
- ❑ Model radio control Rx/Tx set
- ❑ Heavy weight
- ❑ Batteries / battery box
- ❑ Wood
- ❑ Hot-melt glue

What You Do

It's probably best to start by making a one-dimensional (forward-backward) skidmobile. The 1-D skidmobile employs two motors that drive wheels in opposite directions. A third freewheeling wheel provides balance. Four wheels would create problems, because you could not guarantee, except by accurate chassis construction and the use of suspension springs, that all four wheels would always sit properly and evenly weighted on the ground.

I made my chassis out of wood, just gluing the active components onto a central block with hot-melt glue. The servo on the top had a swinging arm along which I could position the operating weight at different distances from the axis. You need to ensure that the entire vehicle has a center of mass that is

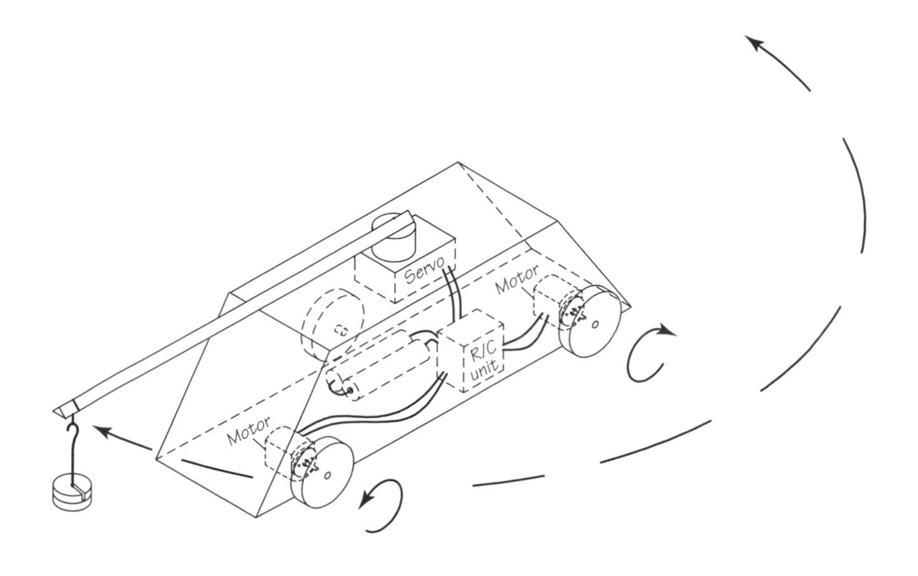

halfway between the driving wheels, but otherwise the placement of the parts is not critical.

The motors I used ran from 3-volt batteries, and each incorporated a simple gearwheel transmission, purchased as a preassembled unit for use in toys, with a reduction of about 100:1. They turned once or twice a second, depending upon how heavily they were loaded. I used smooth polythene plastic wheels for my skidmobile, and ran it on wooden surfaces (benches and tabletops). Its behavior on floors isn't as predictable: some vinyl floors have a rubbery grip on the wheels, and the skidmobile turns and squirms unexpectedly on them.

How It Works

The skidmobile works because of the differential weight on each of the skidding wheels. If the movement of the steering weight caused the entire vehicle's weight to be over the forward-going wheel, then that's the wheel that will propel the vehicle; with much less weight over the backward-going wheel, that wheel skids, which allows the vehicle's forward progress. The situation is reversed with the steering weight over the backward-going wheel. With smaller movements of the steering weight, the situation is not so cut and dried, but the principle stands.

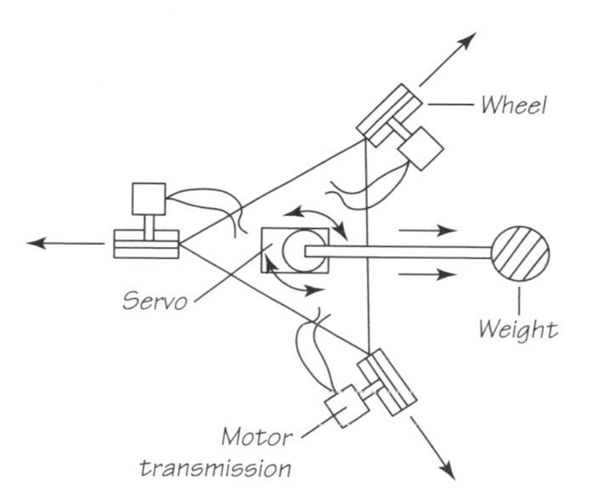

If your 1-D skidmobile works well, you could try making a 2-D skidmobile, capable of moving in any direction. The diagram shows the kind of arrangement you could try. The use of three wheels avoids any issues with alignment of the chassis. The wheels need to point in roughly 120-degree separated directions, and they must all be powered. Once built, place your 2-D skidmobile on a surface where it cannot come to any harm. Unlike the 1-D vehicle, the simple 2-D skidmobile cannot be brought to a halt by command, since its steering weight is always off axis: it is fated to move continuously until it is stopped by an obstacle, runs out of battery power, or falls off the tabletop.

THE SCIENCE AND THE MATH

The top speed of a skidmobile is determined by the speed of its wheels. Once it finishes accelerating, it moves along with the trailing wheel skidding 100 percent, while the leading wheel is virtually skid-free. As you will discover, however, a typical skidmobile takes quite a while to get near that top speed. To understand why, examine the law of friction.

Approximate though it may be, the Newtonian law of friction provides a reasonable explanation for how the skidmobile behaves, at least at low speeds. The approximate law of friction declares that the frictional force F acting against motion on an object dragged across a plane surface depends only upon the natures of the object and plane surface, which are incorporated in a single coefficient of friction μ and the force perpendicular to the plane F_p:

$$F = \mu F_p.$$

In the case of the skidmobile, we have two opposing forces arising from different perpendicular forces $F_{p1} - F_{p2}$ acting on the wheels at either end, giving rise to a net force ΔF:

$$\Delta F = \mu(F_{p1} - F_{p2}).$$

If we assume that M, the vehicle mass minus the moving weight, is divided equally among the three wheels, and that the increase in force over the wheel due to the moving weight is mg, then we have

$$\Delta F = \mu([M/3 + m] - [M/3])g$$

$$\Delta F = \mu mg.$$

The resultant acceleration can be approximated by the Newtonian formula $F = ma$:

$$a = \mu mg/M.$$

For my demonstration machine, I used a 25-g (1-ounce) moving weight, and the whole machine minus that weight weighed about a half pound (240 g). I estimated the coefficient of friction μ to be about 0.13 by measuring the minimum slope angle θ_{min} at which the machine slid freely ($\mu = \tan\theta_{min}$). The anticipated acceleration would therefore be about 0.13 ms^{-2}, or about $g/75$. This seems to accord well with the demonstrations of the skidmobile, which takes several seconds to move across a tabletop 1.5 m (5 feet) in diameter.

At higher speed, other considerations come into play. For example, slight unevenness in the wheel will allow the wheels to jump momentarily from the surface, reducing the average frictional force. So increasing the skidmobile's motor power can actually *decrease* its acceleration, although increasing the power will increase its final top speed.

ANTISKID TECHNOLOGY

The manufacturers of cars and trucks spend huge sums on research and development to prevent skidding: antiskid brake systems (ABS) are nearly ubiquitous now, along with anti-jackknife systems for trucks, brake-pressure regulators, traction control, and differential braking-steering systems. Years ago, cars incorporated simple regulators controlled by suspension position: when a car tipped nose-down, the hydraulic pressure on the rear-wheel brakes was reduced. Today the simplest system is probably ABS, a control system for hydraulic brakes that, when it detects skidding on a wheel, leaks hydraulic fluid from the brake line to that wheel and reduces the hydraulic pressure. With lower pressure, the wheel is less braked and almost instantly begins to roll freely again. When the wheel is rolling, the sensor allows hydraulic pressure to build up and brake the wheel again. The sensor varies among vehicles, but the most ingenious system uses a simple magnetic-field detector, arranged with a magnet behind it next to the shaft on which the wheel is mounted. With a perfectly cylindrical shaft, such a system would produce no signal at all. However, with a set of grooves on the shaft, the magnetic sensor produces pulses as the shaft rotates, pulses that stop if the wheel is overbraked and skidding. The cessation or slowing of the pulse stream activates the hydraulic pressure relief valve.

And Finally . . . the Lubomotive

Perhaps you can devise a simple means of controlling both the speed and direction of a 2-D skidmobile. You will need to vary not just the bearing direction of the steering weight, but also the amount by which it is off-axis. To use mathspeak, you need to control the radius and azimuth. To do so you could fit another servo—a "speed" servo—to the rotating arm of the steering servo. The speed servo would move the weight in and out along a horizontal radius using concentric tubes like a trombone, while the steering servo would determine the azimuth. You'll need some mechanical ingenuity to make it all work. A simpler approach might be to use the speed servo to lift up a weight in a vertical plane arc, rather than move it in and out; this may provide a neater solution than trying to make a trombone-style radial movement using the servo.

There may be other ways in which a skidmobile could be operated. Another Marx Brothers–style vehicle, for example, could apply lubricant differentially to its trailing wheels, leaving the leading wheels to grip well. Maybe that would have to be called a lubomotive rather than a skidmobile. A lubomotive would have to use a plentiful, inexpensive, and easily cleaned-up lubricant, of course. Gelatin-based "slime" could be a possibility, but maybe plain water would be enough.

The opposite effect might be used too: negative lubricant, aka sand. Old-fashioned steam locomotives on railroads used to use sand-dispensing boxes, for example. But with a skidmobile arrangement, a few moments after one applied sand to the leading wheel, the trailing wheels might suddenly wrest power back again from the leading wheel. A lubomotive wouldn't suffer from that problem, because its leading wheels wouldn't be contaminated by the lubricant applied to the trailing wheels. And I suspect that sand would be harder to handle than water. Two or three small pumps, a tank of water, and a radio control system to switch the pumps on and off may be all that you need for a lubomotive engine. Of course, a lubomotive may run erratically if it drives around in a loop and thus accidentally runs over its own trail of slime!

Curious Controls

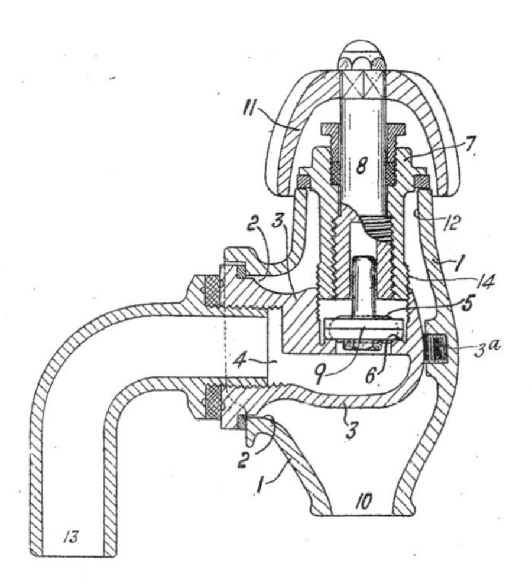

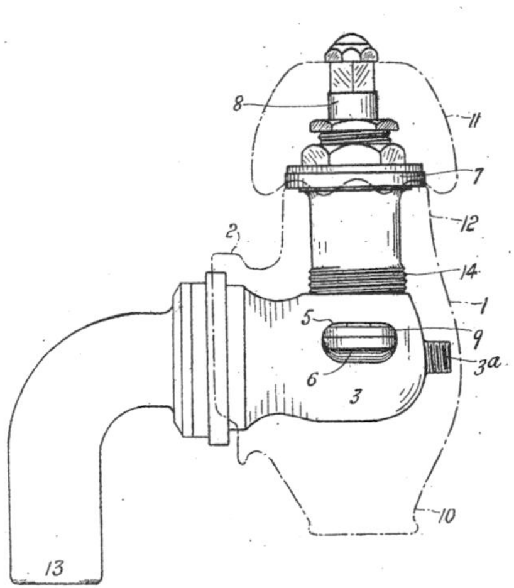

It has long been clear to me that the modern ultra-rapid computing machine was in principle an ideal central nervous system . . . that its input and output . . . might be artificial sense organs . . . and motors or solenoids. With the aid of strain gauges to "feed back" to the central control system . . . we are in a position to construct artificial machines. . . . It gives the human race a new and most effective collection of mechanical slaves to perform its labor.

—Norbert Wiener, *Cybernetics; or, Control and Communication in the Animal and the Machine* (1947)

In my work, I need to control the flow of gases, and the only way to do this in most systems of gas pipes and vessels is to use a valve. Valves come in various shapes and sizes, but practically all of them, in effect, move a seal across a hole in the pipe. The geometry of the seal varies between a sliding action in rotating disks or balls and the clamping seal of the classic faucet-style seat valve. For automatic valves, the pressure of air or hydraulic oil can be used to power a valve. More often, electricity is fed to a solenoid, or, rarely, to an electric motor with reduction drive. These technologies have stood the test of time in a mature industry. The nascent technology of microelectromechanical systems (MEMS) promises to bring a breath of fresh air into the world of gas-flow control. Valves powered by the expansion of liquid, by electrostatic force, or by the momentary boiling of liquids are all technologies that hold much promise in the microworld.

In this section we first explore a flow control that uses none of these approaches. Instead, it exploits one of the curious properties of liquids: the surprisingly strong "surface skin" effect we call surface tension. We then review an electromechanical control that takes a similarly maverick approach to provide a variable speed control of a rotating shaft, discarding conventional ideas like brake pads and using an electric current to create a variable amount of lubricant. Finally, we assemble a gadget that could have revolutionized nineteenth-century science and technology: the piezistor. The piezistor is my name for a valve that uses a piezoelectric device to control the flow of electricity, rather like a transistor. The piezistor is much more limited in performance than a transistor, but it could have been constructed sixty or seventy years before transistor technology came along. Would the piezistor have changed the course of science—or the course of world history more generally—had it been discovered in Victorian times?

14 Wet Blanket Valve Flow Control

Big whirls have little whirls,
That feed on their velocity;
And little whirls have lesser whirls,
And so on to viscosity.

—Lewis Richardson, in his 1920 paper on turbulence,
"The Supply of Energy from and to Atmospheric Eddies"*

Even an apparently simple material such as tissue paper has many different properties that define its behavior. Like Lewis Richardson's turbulence, tissue exhibits structure at many scales, ranging from millimeter-size patches through bunches of fibers to individual fibers made up of myriad tiny internal tubules only nanometers across. Look at a pieces of tissue with a magnifying lens, then with a microscope, and you will see this. Tissue strength, thickness, absorbency, color, feel, and other properties are all important and depend upon the microstructure of the material. Special properties may be important too: tissues used in the clean rooms of the electronics industry, for example, must not shed fibers, even to the most minute degree.

Because of such requirements, paper manufacturers have developed considerable expertise and a considerable amount of instrumentation devoted to controlling their products. One of the instruments used is a "ball burster," which measures the strength of tissue paper when wet or dry. Another instrument, used by filter-paper makers, measures the "bubble point," the pressure that must be

*Richardson was echoing Jonathan Swift's original: "So naturalists observe, a flea / Hath smaller fleas that on him prey; / And these have smaller fleas to bite 'em / And so proceed ad infinitem."

exerted to blow air through a wet filter. Water or another liquid absorbed into the paper will block up pores in the tissue, which reduces or stops airflow. It is this effect that we put to work in the wet blanket valve flow control.

What You Need

- ❑ Fan
- ❑ Tube to fit fan
- ❑ Mesh grid to fit tube (if necessary)
- ❑ Tissue or other highly porous nonwoven material
- ❑ Water spray gun (e.g., plant sprayer)
- ❑ Flow meter (e.g., rotating-vane type; see text description)
- ❑ Funnel on which to mount flow meter

What You Do

The diagram shows the features required for the project: a fan blows air through a tube that has a sheet of material blocking it, a sheet of tissue or cloth that can be wetted by pumping a spray. The airflow is monitored by a rotating-vane flow meter.

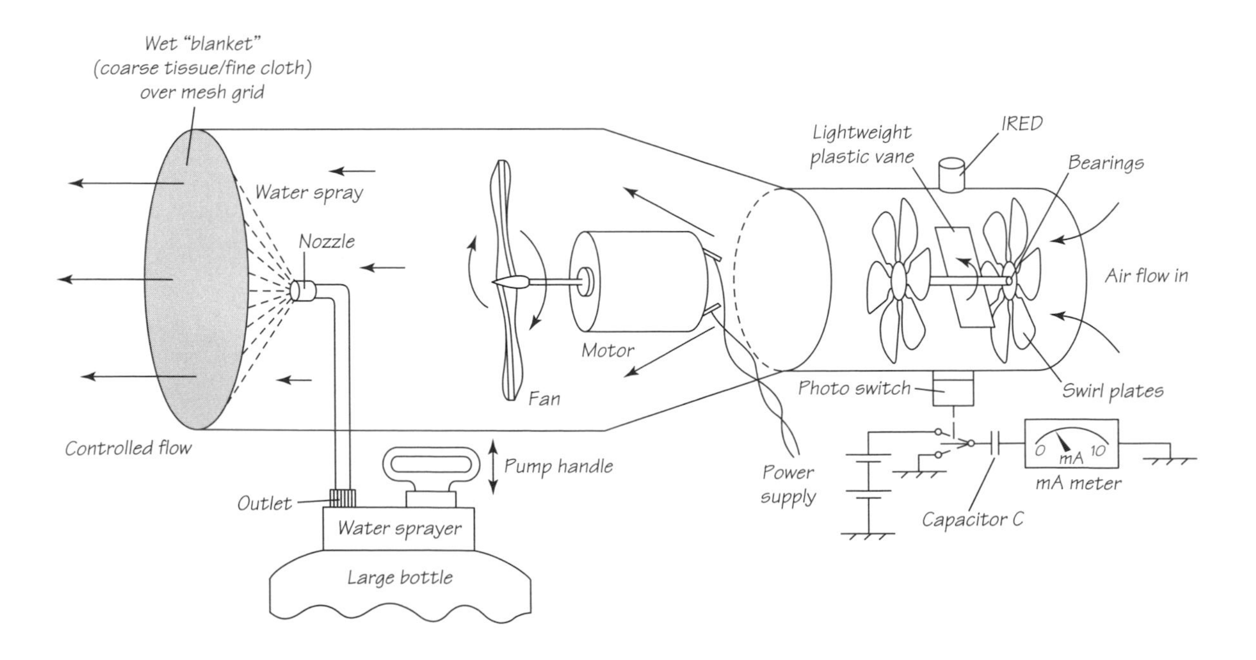

I used an electric cooling fan, of the type used to cool larger computers and other electronics equipment. The plant sprayer must be positioned to inject a fine mist on the back of the tissue or cloth. The mesh grid supports the tissue so that it will not tear too easily when wet. I use the term "tissue" to cover the large variety of nonwoven permeable materials sold in stores for various cleaning purposes. Some type of nonwoven wiping cloth is best, because it is stronger than standard handkerchief tissue but still very permeable. With this kind of material, you may not need a support grid, even with a really powerful fan.

The airflow meter I used was a rotating-vane meter of the type I call a "swirl meter." The air down the tube is rotated by the blades of what resembles a stationary turbine disk. This rotating air is then sensed by a simple flat vane that rotates freely on an axis along its middle, which is coaxial with the tube. The number of rotations of the vane will reflect fairly accurately the amount of flow that has passed it—a flat vane is often better at measuring flow than a twisted vane or propeller shape. The rotation of the vane can be gauged visually, or electronically by using a magnet/reed switch, LED/photodiode, or IRED/photodiode arrangement.

Once you have a setup resembling that in the diagram, switch it on and try it. With the fan on, you should get a steady airflow through the tissue. I measured the flow by counting the number of pulses from the swirl meter over a known time. Now pump the spray a number of times, working right up to the point until the tissue is saturated over a few minutes. As you saturate the tissue, you should see the flow rate stepping downward. You will also notice that the tissue dries off rapidly, so that if you stop spraying for a couple of minutes, the airflow rate will begin to climb back up.

How It Works

Once the tissue is substantially wet, the flow is reduced vastly. In my case, it went to less than 10 percent of the flow rate with dry tissue. Can you relate the rate of water added to the tissue with the airflow rate in any way? Is there a linear relation between the water added and the airflow rate reduction? Or is the effect more related to the area that is wet?

Materials differ in their performance; large numbers of pinholes will limit the turndown ratio (maximum flow to minimum flow) to lower values. Instead of nonwoven materials, you could try fine-woven cloths, for example, which should show a smaller variation in hole size and may work well. Beware of cloths that feature rather large holes in the weave, however, even though they may be woven from fine fibers.

If you want to demonstrate a higher flow-rate capability, you could try taking a leaf out of the filter manufacturers' book. Filters often incorporate a pleated surface to increase the surface exposed to flow without increasing the unit size. This strategy would boost the possible maximum flow rates achievable and minimize the pressure-drop effect of the control surface on the fan's air output.

THE SCIENCE AND THE MATH

Why does wet tissue paper stop the flow of air? Basically, the water fills up the pores and is held in place by its own elastic "skin," formed by the force we call surface tension. How strong is this skin? We can gauge its strength by means of the "bubble point" concept. A filter's bubble point is the pressure at which air will push bubbles out of the filter when it is saturated with water. Given a standard wetting liquid, the bubble point characterizes the maximum pore size of a filter, which we can understand if we consider both the gas pressure on a bubble and the surface tension within it.

Visualize a spherical bubble of radius R cut in half. The force F on the diametral surface will simply be the internal pressure P multiplied by the area of the circle, πR^2:

$$F = P\pi R^2.$$

But this force is also the tension force in the cut-bubble material, which is given by T, the surface tension along the cut, multiplied by the length of the circle, $2\pi R$:

$$F = T \cdot 2\pi R.$$

Hence $T = PR/2$, and thus

$$P_{bp} = 2T/R.$$

The bubble point pressure P_{bp} is proportional to the surface tension and inversely proportional to the bubble radius, which is approximately the maximum pore size of the material. For water, with $T =$ 73 mN/m (a relatively high value compared with organic solvents, for example) and with a maximum pore size of 200 microns, we can expect

$$P_{bp} = 2 \cdot 73 \cdot 10^{-3}/2 \times 10^{-4} = 730 \text{ Pa or } 7 \text{ mbarg}.$$

This is the kind of pressure differential produced by an ordinary electric fan. Because there will be at least a few holes larger than this maximum pore size, the tissue will allow a small airflow even when fully soaked. Also, random turbulent flow from the fan will evaporate some of the water. Once a little evaporation has removed water from the tissue, flow will resume and will allow more evaporation, quickly clearing some of the tissue and enabling the "valve" to open.

And Finally . . . the Trouble with Soap

What is the effect of soap? Soaps and other surfactant molecules lower the water surface tension and reduce the bubble-point pressure. Soap is thus one reason why I suggest that you do not try starting with cloth. Unless it has been especially carefully rinsed, cloth often contains a bit of soap, fabric conditioner, or other material in a quantity sufficient to reduce surface tension.

There are other surfactants, however, that can increase the effective surface tension. Does the addition of silicone or another water-repellent material help or hinder the action of the wet blanket valve flow control?

15 Electrolytic Gas Pedal

> Fooling around with alternating current is just a waste of time. Nobody will use it, ever.
>
> —Thomas Edison, 1889

Among the many inventions that Thomas Edison devised was a curious electrolytic loudspeaker that enabled him to offer an early telephone system in competition with Alexander Graham Bell. Bell's devices went on to become the ubiquitous loudspeakers we use today, while Edison's device was forgotten after a few years. But at the time Edison gave Bell a run for his money, which probably speeded up the development of the infant telephone industry. (Edison also championed his own DC electric supply system and often ridiculed the arguments of competitor George Westinghouse in favor of the AC electricity system we use today.)

The Edison loudspeaker, although scratchy and noisy, was much louder than the Bell telephone receiver, which produced a high-quality but almost inaudibly soft sound. Edison's device employed a cylinder of chalk, moistened with salt water, that rotated, rubbing against a diaphragm. Once a direct current had been established alongside the incoming signal current, you could clearly hear, among the loud noises of the cylinder rubbing, a reproduction of the sounds from the distant microphone.

Improbable as it may seem, electrolysis was the phenomenon that made this curious device work. If you pass a fairly large current—a few amps at least—

through salt water, a visible stream of hydrogen, chlorine, and oxygen bubbles will result. A current of a few milliamps will not. And surely switching such a tiny current on and off with the peaks and troughs of the sound waves of voices or music on a millisecond timescale could not possibly produce bubbles "in time to the music" and thus reproduce sound? Surprisingly, it does.

You can demonstrate the Edison loudspeaker effect fairly easily, as we show below, without the aid of any pieces of antique telephone equipment. We then explore how you might use the same effect in another way, to control the speed of a motor.

What You Need

Edison Coffee-Cup Loudspeaker

- ❑ Radio
- ❑ Salt water
- ❑ Piece of paper
- ❑ Metal strip (e.g., aluminum, 0.5 or 1 mm thick)
- ❑ Piece of metal plate (e.g., quarto size [A4], $8\frac{1}{2} \times 11$ inches, or bigger if possible)
- ❑ Disposable plastic cup (preferably not expanded-foam type)
- ❑ Alligator clips
- ❑ Wire
- ❑ Glue

Electrolytic Gas Pedal

- ❑ Motor
- ❑ Switch
- ❑ Batteries
- ❑ Wood for base and support
- ❑ Glue
- ❑ Paper
- ❑ Salt
- ❑ Water
- ❑ Metal strip
- ❑ Hinge
- ❑ Weights
- ❑ Springy metal strip (e.g., a used minihacksaw blade)
- ❑ Wires
- ❑ Alligator clips

Edison Coffee-Cup Loudspeaker: What You Do

You can make a working Edison loudspeaker with little more than a coffee cup and a few wires and battery. The coffee cup is used as a "sounding board," while a piece of bent metal strip acts as a slider whose friction value depends upon the current flowing at audio frequencies. To demonstrate the effect, bend the metal strip into shape, making sure that no sharp edges touch the paper when you slide the metal along it. I used a piece of aluminum about 0.5 mm thick. Glue the metal part carefully—you need a strong joint—onto the middle of the coffee cup's base.

Now soak the paper in salt water, place the paper on the metal sheet, and connect the radio as shown. Current should flow from the battery via the metal strip on the cup through the dampened paper to the metal plate and then back to the battery. Tune the radio to a suitable channel—a music channel may be best—and switch the volume control to maximum, then disconnect the loudspeaker and draw the coffee-cup loudspeaker across the wet paper. You should hear sounds—rather quiet and distorted, but sounds nevertheless. The small electric currents flowing to and fro in time to the vibrations of the music cause electrolysis when the current reaches the salt water. The electrolysis evolves into tiny amounts of gas—minute bubbles—that act as a lubricant and allow the metal piece to slip. This pattern of slipping on a bubble, sticking on the paper, then slipping on a bubble again causes the metal to pull the bottom of the cup in and out, transmitting sound waves into the air: the cup is thus acting as a loudspeaker.

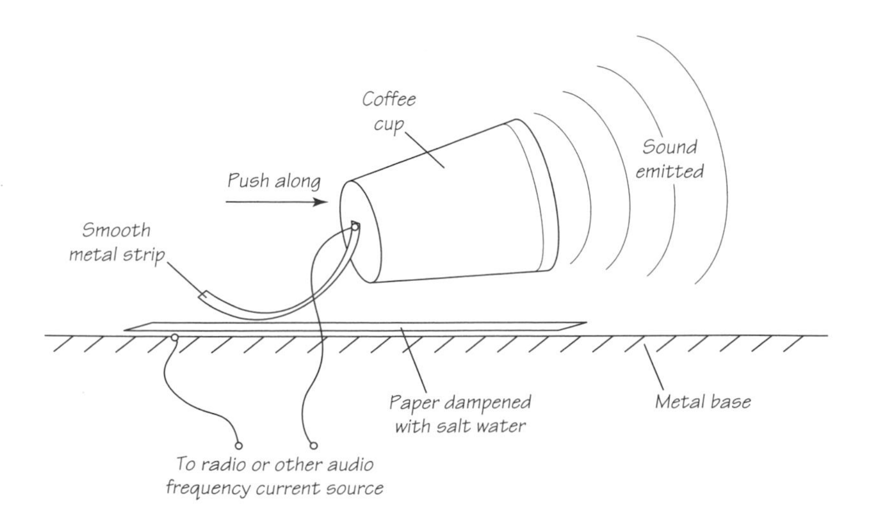

Electrolytic Gas Pedal: What You Do

The diagram shows one way in which you can apply a friction brake to a motor. This is not of itself an unusual arrangement, but the brake we will use *is* unusual: it is porous, and within it bubbles of hydrogen, chlorine, and oxygen gases can be produced to act as a gaseous lubricant.

The arrangement shown is one of two I tried, and many more configurations are possible. There are a number of things to note. First, the spring steel contact (I used a hacksaw blade) on the back of the motor shaft needs to be pushing gently—just hard enough and no more—on the shaft's end to keep an electric contact. Second, take care with the position of the salty paper pad (the brake pad) on the metal strip. The current flows from the battery via the springy steel through the motor shaft to the brake pad, and then through the brake pad, returning via the weighted metal strip and a flexible wire to the battery. The pad needs to be fairly wet but not dripping, and it needs to be fairly heavily doped with salt in order to conduct enough current.

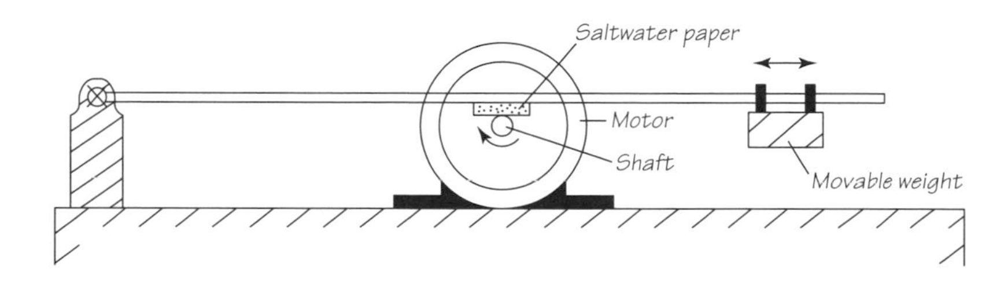

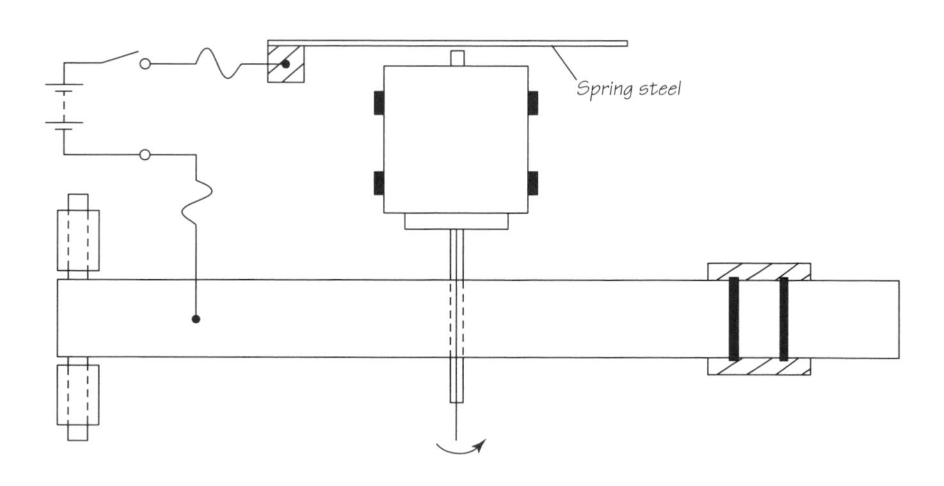

Now switch the electrolysis current on and off. With the motor running at a few thousand rpm, you should find that the motor speeds and slows by at least a few hundred rpm: at 2,000 rpm with a 2-mm shaft and an electrolytic current of about 0.2 A, I found that I could slow the motor down to 1,500 or even 1,000 rpm by cutting the electrolysis, depending upon the weight applied. Just as with the Edison loudspeaker, the motor control effect relies on the production of tiny invisible bubbles of gas, which act as a lubricant.

Further Experiments

What difference does the polarity of the electric current make? How predictable is the speed of the motor when you have electrolysis current connected? How much does the motor speed vary with an additional load—e.g., with your fingers acting as an additional brake on the motor shaft—and electrolysis on and off? Is there a regulator effect—does the electrolysis stabilize the motor speed at all? Would a different salt, or an alkali, work better? An acid would probably attack the steel motor shaft too vigorously. After you have finished testing, always switch off the electrolysis current as well as the motor current, and wash the salt solution off the motor and the other components, because corrosion will set in very quickly.

Try adjusting the size and position of the movable weight on the metal strip. With more torque applied to the metal strip, the motor will be more effectively braked. Beware not to overdo it, as motors tend to run erratically. More seriously, they tend to overheat if you overload them.

Try varying the size of the shaft, if you have motors with different shaft sizes (I found motors with 2-, 3-, and 3.5-mm-diameter shafts), and the power applied to the motor. You need to adjust the voltage supplied to the motor to achieve a reasonably moderate speed (a few hundred or low thousands rpm) with each. A larger motor with a wider shaft will generally run on higher power and may also be intended for higher voltage supply. By applying a lower voltage than intended, you can restrain the amount of power applied to keep it within what the electrolytic gas pedal can control. With luck, by tuning the power supplied, you should be able to demonstrate and measure the same kind of power control with different motors.

MEASURING MOTOR RPM

You may wonder how I measured the rotation speed of the motor. I tried two methods: one that used the intrinsic properties of a commutator motor, and the other that detected the rotation of the shaft more directly. If you connect an oscilloscope probe to the connections of a working electric motor, you will see periodic dips in the voltage supply, as pulses of current flow, following the disconnections and connections made by the commutator. It is easy to measure the pulse width on the oscilloscope to obtain an approximate rpm figure.

My other method was slightly more complicated to apply, but easier in principle. First you need to add a vane, or a wheel with a series of vanes, to the motor shaft. You can find such wheels ready-made inside old computer mice. Now take an infrared diode and infrared phototransistor and glue them onto a U-shaped piece of material so that the diode beam goes into the phototransistor across thc gap in the U. You can also find devices like this inside old computer mice, looking like a squarish U-shaped piece of black resin with four wires coming out of the bottom. You need to mount the LED/diode set so that the vane or the vanes of the wheel will interrupt the LED beam.

Feed the diode with a current of about 10 mA or 20 mA by putting a resistor in series with a battery, such as a 1-k resistor for a 9-V PP3 battery. Connect the two phototransistor leads to the input of an oscilloscope or, better, a multimeter that has a frequency measurement capability. If you now connect a battery—a 1.5-volt or 1.2-volt single cell should do—via a 50-k resistor to the correct terminal of the phototransistor, the multimeter should read the rpm ÷ 60 (assuming that the readout is in Herz).

THE SCIENCE AND THE MATH

The motor's 3-mm shaft, with the motor rotating at about 1,800 rpm, has a peripheral speed of $30.2\pi \times 0.0015 = 0.3\ ms^{-1}$. Electrolysis with salt water yields hydrogen at the negative electrode and chlorine and oxygen at the positive electrode:

Negative electrode: $2H_3O^+ + 2e^- \rightarrow 2H_2O + H_2$

Positive electrode:

$2Cl^- \rightarrow Cl_2 + 2e^-$

and

$4OH^- \rightarrow 2H_2O + O_2 + 4e^-$.

The hydroxonium (H_3O^+) ions require less energy to discharge than the sodium (Na^+) ions from the salt, and so hydrogen gas evolves rather than sodium metal. At first, mostly chlorine is produced at the positive electrode, but as more chloride (Cl^-) ions are discharged, the concentration of hydroxide (OH^-) ions increases and more oxygen is produced.

The volume of gases produced is rather small. It can be calculated using the principles expounded originally by Michael Faraday. Dealing only with the H_2 for the moment, at a current of 0.2 A, we have 0.2 coulombs of electric charge, which amounts to $0.2/1.6 \times 10^{-19}$ electrons, which would produce half as many molecules of H_2. We can estimate the volume of gas produced by noting that 22,400 cm^3 of gas at normal temperature and pressure contain an Avogadro's number (6.022×10^{23}) of molecules:

$22{,}400 \times 0.2/(1.6 \times 10^{-19} \times 6.022 \times 10^{23})$
$= 0.047\ cm^3$.

What precisely happens to the gas formed on the motor shaft "electrode"? If we assume that all the gas formed goes into a thin gas lubricating film, then we can perform some calculations. At the rate of gas production indicated by the Faraday calculation above, if we assume that all the gas generated is moved off the brake surface by a single turn of the motor, then the volume of gas within the surface film will be, at 1,800 rpm or 30 rps, thirty times smaller than the above volume—just $0.0015\ cm^3$. This volume of gas, spread over the pad/shaft contact area—about 2×6 mm—amounts to a gas film that is 0.13 mm thick: quite thick enough to account for the observed reduction in friction.

This calculation also exposes the "regulator" effect of the electrolytic gas film that you may have seen. If the motor turns faster, it drags away the gas film faster, and that increases the braking effect of the pad, which slows it down. But this process in turn allows a thicker gas film to build up, which reduces the braking effect . . . and so on.

And Finally . . . Different Speeds, Other Currents

You could try adding a reduction transmission to the arrangement. The Edison loudspeaker effect works well at linear speeds of around 5 cm per second, which is probably slower than a typical small motor will conveniently run. Quite possibly the electrolytic gas pedal will work better at slower linear speeds.

The Edison loudspeaker produces a distorted sound partly because of the frequency doubling effect: both positive and negative excursions of the audio frequency current produce a slipping (i.e., positive) movement on the coffee-cup diaphragm. So a sine wave of frequency f becomes a "rectified sine" wave of basic frequency $2f$. Could you improve the Edison coffee-cup loudspeaker by using a DC bias current—a steady current that is superimposed on the radio signals? This ought to improve the intelligibility of the output, since it will not undergo the frequency-doubling effect noted above. Something of this kind seems to have been used by Edison in his first telephone system.

REFERENCE

Clark, Ronald W. *Edison: The Man Who Made the Future*. London: Macdonald and Jane, 1977.

16 Piezistor: A Victorian Amplifier

In 1948, Bell Laboratories' scientists found that amplification could be obtained by means of a third contact to the normal p-n contacts in germanium. This was termed a point-contact "transistor." . . . The first point-contact transistor had several limitations: it was noisy, it could not control high amounts of power and it had limited applicability.

—Geoffrey W. A. Dummer,
Electronic Inventions and Discoveries

Civilization today would collapse without amplifiers. An amplifier is the sine qua non, the absolute indispensable part, of every computer, radio, television set, telephone, car, and airplane, and of innumerable devices in medicine and science. Yet it was not always so. In the nineteenth century, computation was carried out only with hand-cranked machines or with pencil and paper. Television did not exist, but vehicles functioned, there were even the beginnings of radio, along with electrocardiograph and other medical equipment and a host of scientific instruments—all without amplifiers. Even global communications did not then need amplifiers. Surprisingly, perhaps, even Victorian long-distance telegraphy did not use amplifiers.

Most of us have forgotten about telegraphy. I sent my last telegram in 1981, between London and Chicago as I recall. Telegraphy spanned oceans and continents because its lower speed enabled the use of small electric signals, and because its digital on/off nature allowed the use of electromagnetic relays to increase the available current at intermediate stations and at the receiver.

It was the telephone's development in the early twentieth century that most directly pointed to the need for amplifiers. Telegraphy had brought communication to the whole world, but it was a limited medium: it allowed only the low-speed transmission of text, in uppercase only and without the benefit of special symbols. Telephony was streets ahead, for it permitted communication in an almost natural way with all the nuances of speech and without the need for a telegraph clerk as an intermediary. But without amplifiers, telephones could not rival the global reach of the telegraph. Relays could not be used to boost telephone signals: relays are not amplifiers, they are on/off devices, and in any case they cannot react at the millisecond speeds that audio signals need. It was only with the advent of the vacuum tube, developed by pioneers such as Lee de Forest and John Fleming, that an amplifier became available to satisfy this need.

But could amplifiers have been invented earlier, in the Victorian period? And were they? I am not sure of the latter: I can't find conclusive references to actual working devices from that era, but maybe a knowledgeable reader can settle that question. I am sure, however, that an electronic amplifier *could* have been made in 1880 or even earlier, because the device we now describe could have been made then. Pierre Curie, later to achieve fame with his wife Marie in the discovery of radioactivity, had already uncovered the basic principles of the piezoelectric effect. And a number of inventors, including Thomas Edison and his team, had worked on the principle of the carbon granule chamber as a "microphone": a device capable of faithfully converting the sound waves of the human voice into a varying electric current.

What You Need

- ❑ Piezoelectric sounder unit, perhaps new, possibly a ringer unit salvaged from an old telephone
- ❑ Clamp (e.g., small G-clamp or similar, ideally with fairly fine-pitch screw)
- ❑ Small brass bolt (to fit in hole on the front of piezoelectric sounder)
- ❑ Input signal generator: radio set or (better) electronic oscillator
- ❑ Carbon rod (e.g., from inexpensive C- or D-size nonalkaline cell granules)
- ❑ Multimeter with AC millivolt meter or (better) data logger plus PC or oscilloscope
- ❑ Miniature alligator clips
- ❑ Wires
- ❑ Loudspeaker (600-ohm) or 600-ohm resistor

What You Do

The basic principle of the piezistor is straightforward: a piezoelectric device, activated by a positive input signal, squeezes a capsule of carbon granules, lowering their resistance. The granules then conduct more current around the output circuit. The opposite happens when a negative input signal is provided. The net effect is that input voltage changes are translated into output voltage changes that are larger, or at least more powerful (at lower impedance), than the input voltage.

You will need suitable carbon granules. I prepared carbon granules from the carbon rod of a battery as follows: I clamped an inexpensive nonalkaline D cell in a vise and then used a screwdriver to bend out the top of the thin steel jacket so that the top of the cell was relatively free to move. I then grasped the button in the middle of the top with a large pair of pipe-wrench pliers and simply pulled the top, with the carbon rod still attached, out of the cell. A little cleanup yielded the carbon rod, which I then filed on a coarse file to provide my carbon granules.

Push the filings into the central hole of the piezo unit until it appears to be full and then glue it inside the clamp, as shown in the diagram. Make sure that the hole is on the axis of the bolt. Now glue the small brass bolt onto the large bolt, again ensuring that it is coaxial. Now screw the large bolt down until it is just a fraction of a millimeter inside the hole on the piezo: this will stop any carbon from escaping.

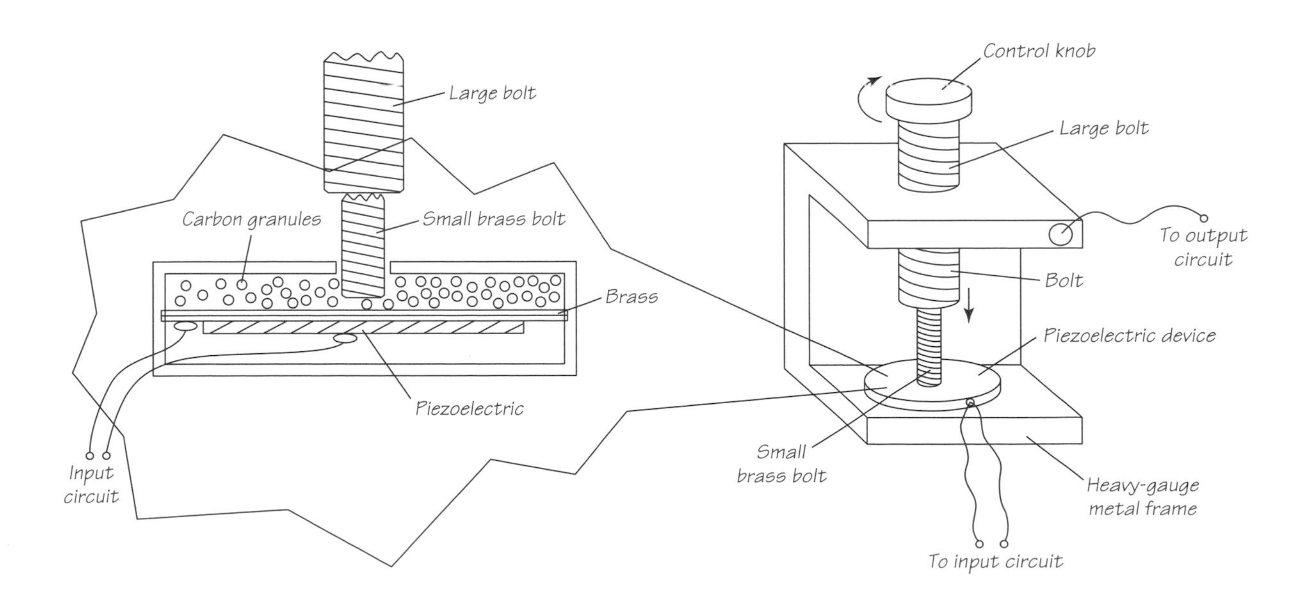

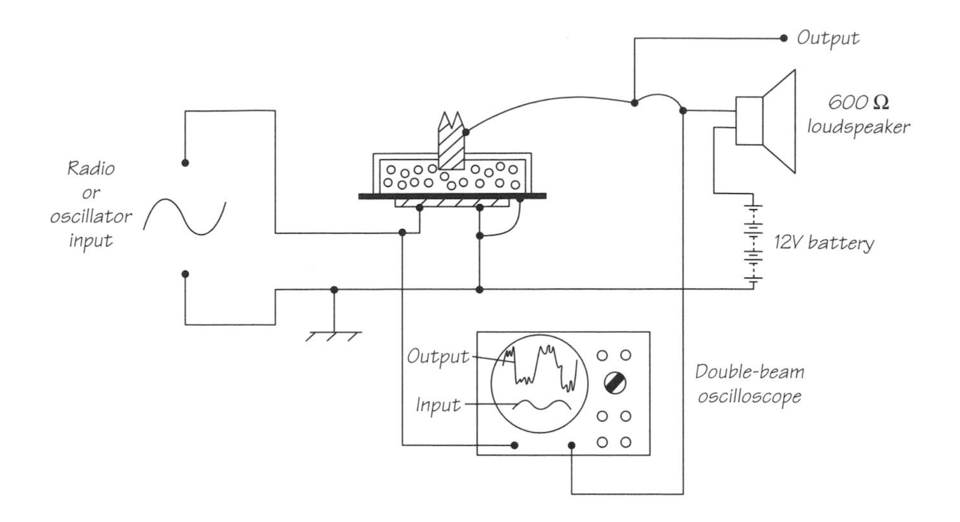

You now need to connect your Victorian amplifier with an input and output. Measuring the ratio of the latter to the former will tell you if your piezistor works. Connect the circuit as shown. A 600-ohm loudspeaker can be used for an audible output, or a simple resistor of a similar value can be used if you only want to monitor the electrical output. Now connect your oscillator, or the radio set, with a volt or two of amplitude being a good starting place. A simple multivibrator oscillator will suffice, running at a few hundred Hz, if you don't have anything else (see the circuit diagram on page 245 for a typical oscillator circuit). You should be able to hear a quiet, muffled sound escaping from piezo.

Connect the 12-volt supply via the 600-ohm loudspeaker or resistor. Then hook up a voltage measurement on the output, to see whether the device is working. You will probably need to adjust the screw, screwing it very slowly down onto the piezo device. Don't overdo it, or you may crack the brittle piezo ceramic, which is on the back of the piece of brass you can see through the hole in the plastic housing. At some point, you should see the voltage drop from 12 volts to some intermediate value; superimposed on this voltage should be a waveform that reproduces the input.

How It Works

As you turn the screw, you should see the waveform change in amplitude. At large amplitudes, you will see "clipping." This occurs when only the peaks of the input waveform are reproduced, with the troughs being cut off, or vice versa. Try adjusting the screw so that the waveform is complete. You will also see noise spikes, but the basic oscillator or radio waveform should make it through. If you have a two-channel data logger or oscilloscope, set it up so that you can see input and output simultaneously and check the output waveform against the input.

If you are listening to a loudspeaker output, you can adjust the device for maximum loudness; if listening on an oscilloscope, multimeter, or data logger, you can adjust for maximum AC voltage. The radio sounds will be a bit distorted, muffled, and noisy relative to a regular radio, but perfectly recognizable. If you are having trouble, check that the carbon granules are still in place, and that you have put enough into the device—just undo the screw and carefully shuffle a bit more carbon into the piezo unit if it seems to be insufficiently filled.

THE SCIENCE AND THE MATH

The piezoelectric effect arises because of the internal geometry of the arrays of charged ions that lie inside some crystals. Their geometry is such that, when a voltage is applied, the electric field makes those ions move, causing a distortion of the crystal lattice, which we see as a small change in one or other of the crystal's dimensions. Basic solid-state physics books (such as Kittel, *Introduction to Solid State Physics*) explain this process, while specialist books (such as Wang, Herbert, and Glass, *Applications of Ferroelectric Polymers*) give all the details of the piezoelectric effect (and the related ferroelectric effect) and explain how they can be used in devices.

A piezoelectric sounder typically consists of a circular piece of thin brass sheet, held by its edges in a housing with a smallish hole in the middle of a plastic housing. A disk of piezoelectric ceramic sheet is stuck to the hidden inner side of the brass. When voltage is applied to the ceramic, it contracts slightly, while the brass does not; this bends the unit. As it moves forward or backward relative to its mounting, the unit emits sound via the hole in the housing. Because of the thinness of the sheets, the geometry is such that even a small (micron) contraction of the ceramic results in a much larger movement of the brass sheet. As shown in the cross-sectional diagram, we have, for the reduction in length t of the ceramic,

$$t = d\Delta R/R$$

where d is the diameter of the sheets that make up the device, ΔR is the difference in mean radius of the two sheets (i.e., the thickness of one of them, if they have the same thickness), and R is the radius of the assembly when bent.

We can also derive, via Pythagoras's theorem, a formula for the deflection A of the sheet from its flat position, when curved to radius R:

$$A \sim d^2/8R.$$

Putting the two together gives us

$$A = td/8\Delta R \text{ or}$$

$$A = K_p V d/8\Delta R,$$

which expresses the deflection in terms of the voltage applied V and assumes a linear contraction, in other words, that $t = K_p V$, where K_p is a constant. To try some numbers, suppose that the ceramic con-

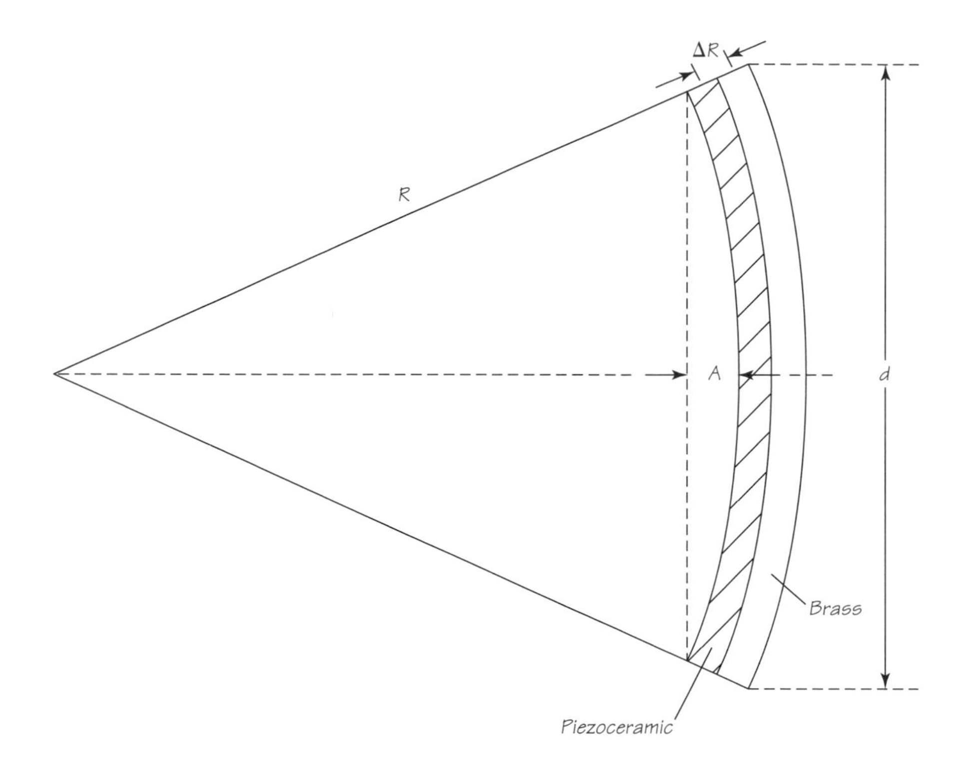

tracts 1 micron over a 20-mm diameter plate, and that it and the brass sheet are just 100 microns thick. Then the assembly will deflect 25 microns.

Just a few tens of microns' deflection is all that is required to change the resistance of the set of carbon granules in the device and thus provide an output. As the resistance of the granules changes, it modulates the output current from the battery through the loudspeaker. It also forms a voltage divider with the load resistor or the loudspeaker and gives a voltage output.

An amplifier is only an amplifier if it amplifies: so a key parameter is its gain factor, which we might define as the ratio of output over input voltage, or output over input power. With the first device I tried, the voltage gain factor was less than 1: the output voltage was less than the input voltage! But the power gain of an amplifier is what we are really interested in. If we have power gain, we should be able to convert it to voltage gain with the aid of a passive device like a transformer, at least in principle. If we don't have power gain, however, then we are in trouble: we really don't have an amplifier in any normal sense of the word.

Input power to the piezistor is not quite so easy to calculate, because it does not have a resistive input impedance: piezoelectric devices at low frequencies are largely capacitive. This being so, the input current I_i and input voltage V_i will not be in phase, and hence we cannot calculate average AC power with the simple relation $P_i = V_iI_i$, where V_i and I_i are the root-mean-square (rms) values of AC voltage and current, respectively. In fact V_iI_i is the maximum power, assuming an in-phase relationship. If we assume for the moment a 50 percent "power factor," in other words that

$$P_i = 0.5V_iI_i,$$

then we can calculate the power gain of the device by calculating the current and voltage flowing in the piezistor. If its capacitance is C, then the current flowing in the input circuit I_i is

$$2\pi f C V_i.$$

The power given is

$$P_i = V_i^2 \pi f C.$$

The output power is $P_o = V_o I_o = V_o^2/R_o$, where R_o is the output load, which we can take to be that of the resistor or the loudspeaker, thus expressing the power gain G_p as

$$G_p = V_o^2/(\pi f C R_o V_i^2).$$

This formula thus allows you to measure the two voltage amplitudes V_o and V_i and to convert these into a power gain. As usual with power gain factors, G_p is inversely proportional to the load resistance: with a low resistance load, the power gain will be higher. You might think, considering this formula, that lowering R_o would lead to a higher power gain. If you try this, however, you will find that the voltage V_o shrinks, becoming inversely proportional to the resistance R_o at low R_o values and removing any advantage. It is in fact typical of most electronic amplifier systems that they show their maximum performance when the output impedance is equal to their load impedance. This is true, for example, when extracting the maximum power from a battery, voltage V, of internal impedance r into a load resistor: you will get the most power out P_{out} when you choose a resistor R which is equal to r:

$$P_{out} = V^2R/(r + R)^2$$

$$dP_{out}/dR = -2(V^2R/[r + R]^3) + (V^2/[r + R]^2)$$

$dP_{out}/dR = 0$ when $R = r$, which is a maximum value for P_{out}.

The power gain formula thus correctly shows how the gain factor will be higher at low frequencies. My piezo device had a capacitance of 16 nanofarads (nF) and yielded about 0.6 volts with a 2-volt input. At 200 Hz my piezistor showed a good gain factor—around ×15. At 2,000 Hz, however, it was only roughly breaking even, at least on the assumption we made on power factor, and it was useful as an amplifier only below 2,000 Hz.

And Finally . . . a Moving-Coil Amplifier

How can you increase the gain factor of the piezistor? What about using a piezo device with a lower capacitance C? A smaller device would have a smaller C, so this is a possibility, but a smaller device may also need a higher voltage V_i to produce the same variation in resistance of the carbon granules. What about using a transformer to obtain a higher voltage onto the piezo? You could even use a piezo device to do the transformation: piezoelectric transformers are just beginning to become widely available. They feature a low-voltage piezoelectric emitter closely coupled to a higher-voltage piezoelectric "microphone." Like a more conventional magnetic transformer, they do not amplify power, but instead convert it from low voltage/high current to high voltage/low current or vice versa. Surprisingly, such an arrangement can achieve high efficiency—90 percent or more—by operating at the resonant frequency of the assembly, although the power levels that can be transformed are, so far, much lower than for magnetic transformers. Wang, Herbert, and Glass's *Applications of Ferroelectric Polymers* describes this idea briefly and provides further references.

You could try a completely different input device for varying the pressure on the carbon granules. In Victorian times the most obvious device would probably have been a magnetic diaphragm/magnet telephone unit like that developed by Alexander Graham Bell. Today it is easier to find a small moving-coil loudspeaker: a 600-ohm type would be an excellent match to the output of a piezistor amplifier unit in a multistage amplifier system.

REFERENCES

Kittel, Charles. *Introduction to Solid State Physics.* 5th ed. New York: Wiley, 1976.

Wang, T. T., J. M. Herbert, and A. M. Glass, eds. *The Applications of Ferroelectric Polymers.* Glasgow: Blackie, 1988.

Granules and Particules

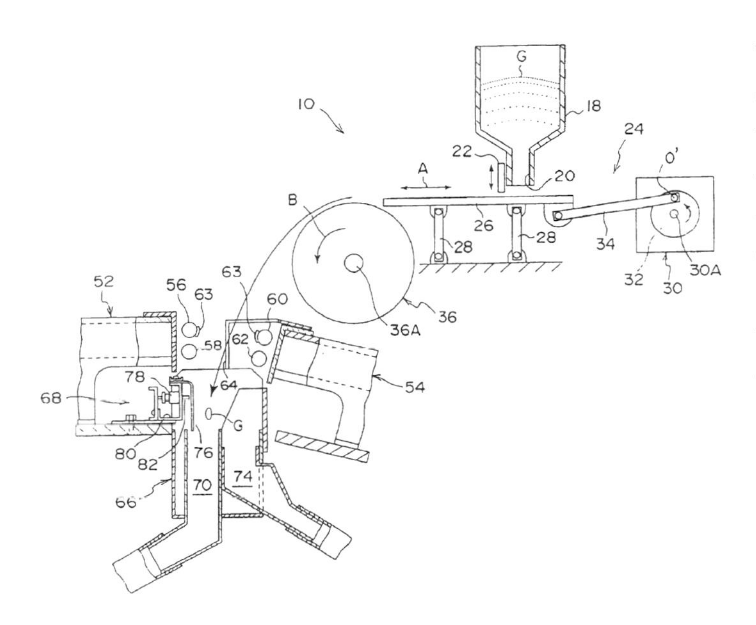

When the Son of man shall come in his glory, and all the holy angels with him, then shall he sit upon the throne of his glory: And before him shall be gathered all nations; and he shall separate them one from another, as a shepherd divideth his sheep from the goats: And he shall set the sheep on his right hand, but the goats on the left.

—Matthew 25:31–33 (King James Version)

Sorting of things—whether animals or objects—is important. One of the first projects I encountered when I began working in industry was a system for separating pieces of crushed mineral. It used a small blast of compressed air to kick out valuable pieces of mineral from a huge stream of unwanted stuff, the useless gangue whizzing past on a conveyor belt. The separation of particles has involved a lot of effort over the centuries, and it remains of considerable importance in industry. Back in the days of the forty-niners, fortunes could be won or lost by picking out a few tiny yellow grains from among a lot of other tiny yellow grains. The more valuable grains, of course, were gold.

More important in the grand scheme of things, however, is the separation of wheat from chaff in farming. This task has been accomplished with the aid of helpful breezes and much manual effort since the earliest days of agriculture, hundreds of millennia ago. It was one of a small number of key activities that enabled the development of agriculture and, ultimately, civilization. And although the machinery used may have changed, wheat is still separated from chaff today.

Similarly, controlling and measuring flows of particulates is important in commerce, although this process has come to the fore more recently, with the advent of large-scale modern farming and mining technologies. Much of today's industry runs with continuous rather than batch processes, and old-fashioned batch weighing or volumetric measure filling, for example, are now no longer good enough. Particulate flow must be constantly controlled and measured, round the clock.

In this section's projects we examine four novel ways of sorting and controlling particulates from seeds to sand, using their physical characteristics: rolling speed, shadow size, magnetism, and aerodynamic drag coefficient, respectively. Who knows? Perhaps one of these projects may even provide the kernel (pun intended) for a new technique that can be put to practical use in some global industry.

17 Rice Grain Ski Jump

> "Sorting" . . . for computer programmers . . . [means] sorting things into ascending order. . . . the process should perhaps be called *ordering,* not sorting; but anyone who tries to call it "ordering" is soon led into confusion. . . . Consider the following sentence, for example: "Since only two of our tape drives were in working order, I was ordered to order more tape units in short order, in order to order the data several orders of magnitude faster." . . . So we find that the word "order" can lead to chaos.
>
> —Donald E. Knuth, *The Art of Computer Programming,* vol. 3, *Sorting and Searching*

Some separations of particulates can be achieved with the simplest possible apparatus. A sieve is not difficult to make, and it works in an obvious way: big lumps don't go through the holes, little pieces do. In favorable circumstances, paradoxically, particulates will separate if they are stirred or shaken around without a sieve, or, indeed, any apparatus at all. If you shake a container of mixed nuts, for example, the Brazil nuts will always come to the surface.

Another unusual separation will occur in a horizontal rotating drum partially filled with a mixture of round particulates. With the correct design parameters, this apparatus can separate particulates whose average characteristics are surprisingly similar. The theory of how this works is complex, but it is tied up with the slope formed by particles of different sizes when they sit in a heap (see Hill and Kakalios, "Reversible Axial Segregation").

The ability to roll efficiently depends upon friction, roundness, and slope. We can separate rice grains in our project here because of their differing propensity to roll and slide down a smooth but still frictional slope.

What You Need

- ❑ Flat-bottomed, U-shaped channel (e.g., roof guttering), 100 mm (4 inches) wide
- ❑ Adjustable mountings for channel (e.g., piles of bricks or books)
- ❑ 2 or 3 small cuboid open-top containers
- ❑ Particles to separate (e.g., round-grain and long-grain rice)
- ❑ Coloring (to color one or both of the rice grains), if desired
- ❑ 2 rigid plastic sheets (about 100 × 40 × 0.5 mm and 150 × 150 × 0.5 mm)

What You Do

The U channel makes a slope for the particles to slide or roll down, and it needs to be adjusted to a suitable angle, which you can determine with a few preliminary trials. The placement of the collecting containers at the bottom of the slope is also subject to experiment. You should position the small piece of rigid plastic sheet to form a short "ski jump" (about 12 mm, or ½ inch) at the end of the sloping channel. The larger piece of plastic, or perhaps a similar size sheet of

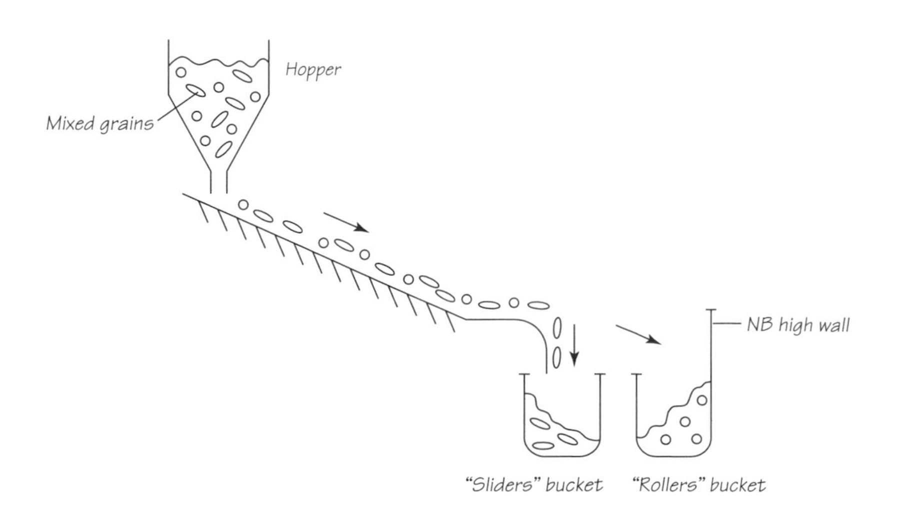

cardboard, should be used to provide a backboard that will stop any particles from overshooting the last container.

I have often found it convenient to prop up the U channel using sheets of paper, sometimes just pages from an opened book. Each sheet of paper will raise whatever it is you are supporting by a precise, small amount, around nine sheets to the millimeter, depending on the paper thickness. Academic and glossy illustrated books seem to provide more precise adjustments, because they don't squeeze down as much under load as do inexpensive paperbacks.

Set up your slope and dribble mixed rice grains from your hand onto the slope, moving the collecting containers into the positions where the different grains—the sliders and the rollers—seem to predominate. Round grains will roll faster down the slope than long grains will slide. When they get to the bottom, the faster grains will follow a parabolic path through the air, whereas the slow grains will fall more or less straight down when they reach the bottom.

What kind of separation can you achieve? As a figure of merit you could use, for example, the ratio of wanted grains to unwanted grains as the separation factor. With ten wanted long grains against a thousand unwanted grains, the separation factor would be 100, for example. You can count the unwanted grains and estimate the wanted grains by weighing. What is the efficiency of this process? What percentage of wanted grains are wasted by falling into the wrong container or no container at all?

The separation is almost nonexistent if you put large handfuls of rice on the slope. With such big heaps, much of the rice will slide, whether round or long. At high densities, many of the long grains bump into the fast-rolling round grains and so speed up. Meanwhile, the round grains lose speed by colliding with the long grains. However, if you dribble the grains in a steady, thin stream, you will find that, at least in the last 150 mm (6 inches) or more of the channel, the grains will typically roll freely, and separation can be almost perfect.

How It Works

A perfectly round particle has a strong tendency to roll on any slope whose coefficient of friction is reasonable. A significantly nonround particle, however, may not roll at all and simply slide when the slope is steep enough. On a steep slope of low friction (such as ice) almost all particles will slide.

There is a further subtlety regarding the nature of roundness: both prolate and oblate ellipsoids can roll, but they do differ markedly in behavior. Prolate

(American or rugby football-shaped) spheroids roll almost as easily as true spheres, because their stable rest position on a surface is with a point on the round edge. Oblate (M&Ms- or Smarties-shaped) spheroids don't roll easily if they are more than slightly nonround, because their stable rest position is not a point on the round edge. You can artificially start oblate ellipsoids rolling on their round edges, but they rarely do so on their own if you drop them at random.

THE SCIENCE AND THE MATH

Once they leave the end of the ski jump, the particles follow a parabolic trajectory, with a steeper average slope for the slower particles. The critical minimum speed v_{min} with which a particle must leave the ski jump at height H_s to clear an obstacle height H_o can be estimated by using the equation of motion:

$$v_{horiz} = \text{constant} = v_{min}$$

$$v_{vert} = gt$$

$$y = H_s - \tfrac{1}{2}gt^2,\ x = v_{min}t,$$

$$\text{so } y = H_s - \tfrac{1}{2}gt^2 = H_o.$$

So we arrive at the ski jump equation:

$$v_{min} = L\sqrt{(g/(2[H_s - H_o])}$$

where g is gravity and L is the horizontal distance between the ski jump exit and the dividing barrier at height H_o.

With any separation process, separation is not perfect. Typically, product A can be almost 100 percent pure, while product B will remain contaminated with 10 or 20 percent or more of product A. Why should this tend to be the case? Why is the production of two pure products through granular separation so difficult? The reasons relate to statistics. Rice

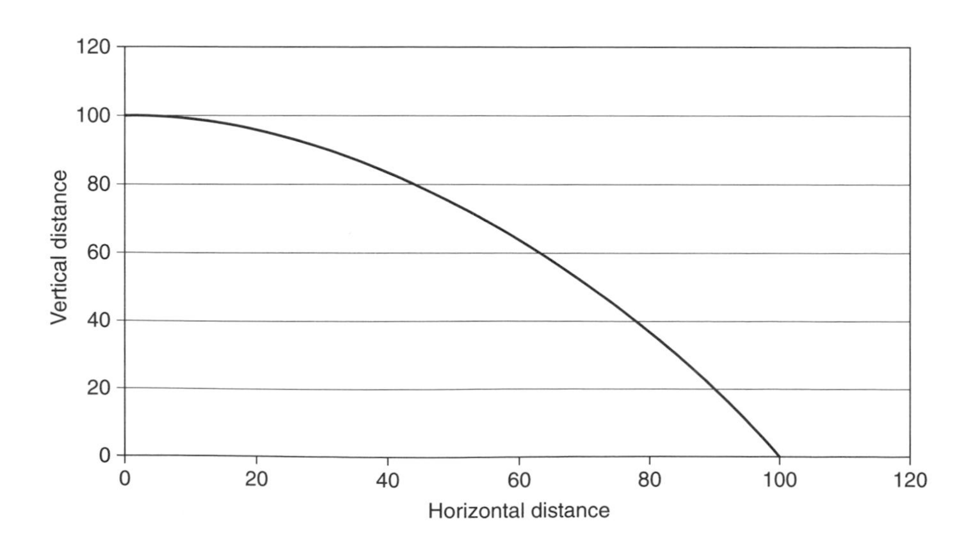

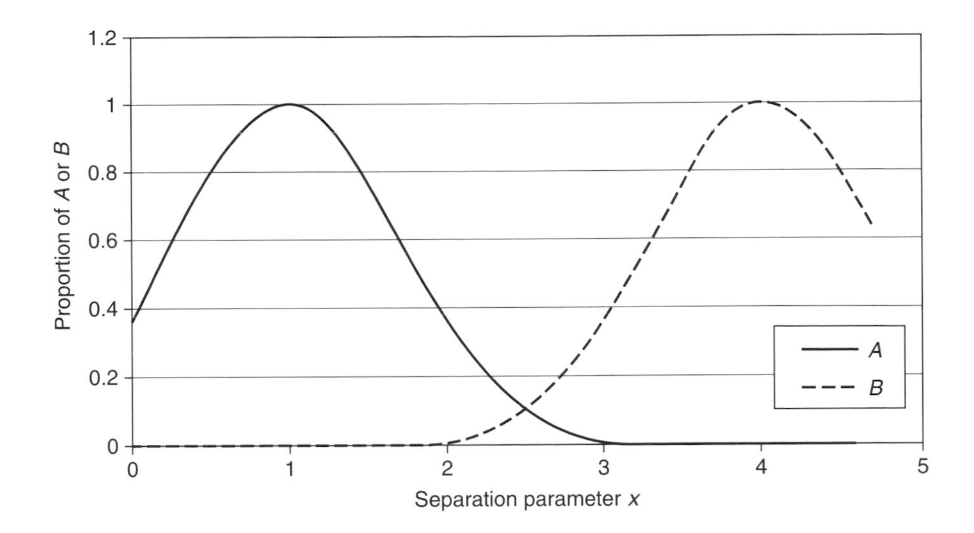

grains are not always identical—there is a spread in their properties. Similarly, the motions along a channel such as the rice grain ski jump are not identical for nominally identical grains. Depending upon random collisions on the way down the channel, some grains will reach the ski jump at a faster speed than others, so there is a spread in takeoff speeds.

In many such situations of random variables, it is approximately correct to consider the spread of parameters of each class of objects to be distributed according to a Gaussian curve, that is,

$$P(x) = P_o \exp(-(x-X_{av})^2/W^2),$$

where X_{av} is the average value of parameter x, and W is a constant parameter controlling the width of the distribution. If we now calculate the area, integrating below the value X_{cutoff} for A, and above the value X_{cutoff} for B, we can obtain the contamination of the two products as a function of the chosen X_{cutoff} value.

An example will illustrate the problem. The graph shows two Gaussian curves, A and B, which have peaks at separation parameter x values 1 and 4, respectively. We could achieve a nearly pure A by establishing a cutoff at 1, that is, ending the A bucket at $x = 1$. But then we would be throwing away half of A, and the other half of A would pollute B, giving B only 67 percent purity. Similarly, we could put in a very wide A bucket, setting the cutoff at 4 and obtaining a 100 percent yield of A, but we would then have purity of only 67 percent A, although B will be pure. If we set the cutoff—the right-hand edge of the A bucket and the left-hand edge of the B bucket—at 2.5, halfway in between, we would have two impure products. Only by throwing away a sizable cut in the middle, from 1.5 to 3.5 perhaps, can we expect reasonably pure A and pure B. But this result comes at the cost of throwing away perhaps 25 percent of each.

And Finally . . .
Pills, Lentils, and Chocolate-Covered Peanuts

What about cylindrical shapes? There will surely be a difference between disk-shaped short cylinders and needle-shaped long cylinders analogous to the preceding discussion of ellipsoids. Maybe you can find some pills that would approximately fit these descriptions. Why not try other separations: between lentils and long-grain rice, or between M&Ms and chocolate-covered peanuts? Which will roll better? Does the rolling motion depend at all upon how you add the grains to the chute? Given that oblate ellipsoids don't roll easily if dropped randomly, can you launch them in a special way so that they will start rolling effectively? How long does the influence of this start persist down the chute?

REFERENCE

Hill, K. M., and J. Kakalios. "Reversible Axial Segregation of Binary Mixtures of Granular Materials." *Physical Review E* 49 (1994): 3610–13.

18 Niagara Meter

> I look forward to the time when the whole water from Lake Erie will find its way to the lower level of Lake Ontario, through machinery, doing more good for the world than that great benefit which we now possess in the contemplation of the splendid scene which we have presented before us at the present by the waterfall. . . . I do not hope that our children's children will ever see the Niagara cataract.
>
> —William Thomson, Lord Kelvin, *Washington Post,* November 18, 1897

A flow meter whose answer depends upon its position along the stream it is measuring is not, at first glance, a reassuring device. But this tool is what you need to measure a flow of material as it streams downward, accelerating under gravity rather than moving at constant velocity. For example, think about the water coming out of a faucet: the appearance of the stream from the nozzle varies with the flow rate, so by careful measurements you ought to be able to estimate the flow rate from the stream size and shape.

You may have noticed how a water stream narrows after it has come out of a nozzle or faucet. Why is this? While the water is inside the cylindrical tail pipe leading to the faucet, it moves along in a cylinder. One would expect that the water's emergence would amount to a removal of the "skin" of this downward-moving cylinder. Surely, therefore, it should just move into the air as a cylinder? Maybe the cylinder should even get fatter as it escapes from the faucet's "skin"—after all, the water in the pipe is under pressure, and surely we are removing this pressure when we remove the "skin." Well, it just ain't so! (See the math analysis for why.)

The Niagara meter uses particles to illustrate aspects of flow measurement. It does not measure the diameter of a stream, but rather its area, and not its cross-sectional area, but rather the sideways projected area of the particles.

What You Need

- ❏ Black plastic drainpipe
- ❏ Piece of black rigid plastic (perhaps cut from pipe) for deflector
- ❏ Photodiode
- ❏ Microamp meter or, if not available, a multimeter with a microamp scale
- ❏ Reading lamp
- ❏ Test particles/grains (e.g., sand, rice, or lentils)
- ❏ Jug to pour out particles in an even stream
- ❏ Translucent cooking paper or tracing paper

What You Do

First you need to build something like the arrangement in the diagram. The idea is that each particle in the stream will partially block the light reaching the photodiode, to an extent proportionate to its projected area and the time that the particle is in the illuminated area. The black tube needs to be long enough that ambient light does not affect things too much. You can try turning the room lights on and off to check that you have shielded ambient light sufficiently.

The photodiode feeds current to the microamp meter or to a multimeter set on its microamp-current range. A multimeter with an analog scale along the bottom will be easier to read than a purely digital readout. While the reading lamp projects bright, even illumination on the translucent screen in the side of the tube, test out the response of the system by putting a stick down the tube in the particle passage area and checking that the multimeter reading goes down. Now adjust the reading lamp so that the meter reads exactly full scale with no blockage of light. Pour some of your test particles through and see what the meter reads.

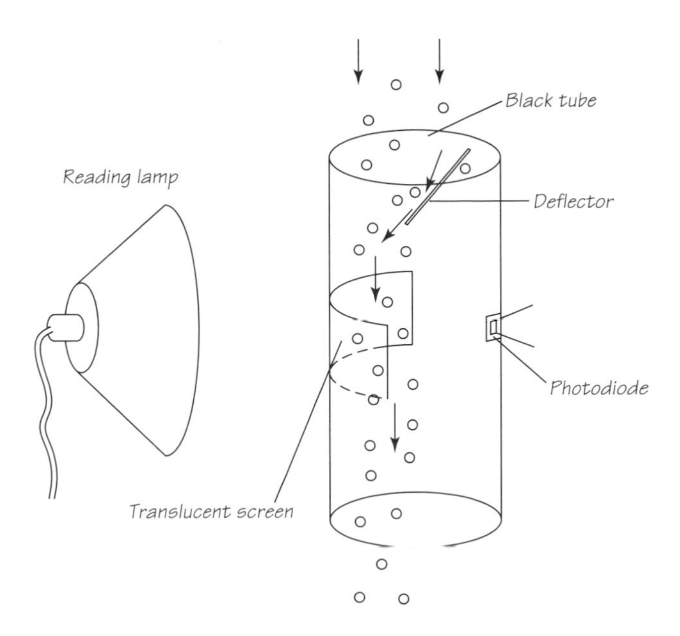

As the particles pour though the tube, the deflector directs the flow to one side to ensure that none drop too close to the photodiode, which would distort measurement. If you are using an analog microamp meter, you could rescale the meter to read from right to left. You should be able to take the cover off and place a paper scale over the face. With a computer drawing program you can draw up a suitable scale with the correct arc of a circle and its gradations.

If you have a data logger device for your computer, try logging the signal during a single pour of particles lasting at least a few tens of seconds. Either draw out the graph and then measure its area, subtracting any baseline present, or add up the readings using the computer. Now repeat this several times, varying your rate of pouring. Are the areas or readings totals equal each time? You will probably find that zero errors mean that the signal is erratic at low rates of pouring, while at high pouring rates you will lose some signal, because some particles will be occulting—hiding behind—other particles.

How It Works

The Niagara meter works because heavy particles accelerate under gravity in exactly the same way as lighter particles. This fact was noted quite a long time ago, most famously by Galileo, who dropped an iron ball and a wood ball off the Leaning Tower of Pisa in Italy to prove the point. (If you visit northwest Italy, be sure to see the Leaning Tower. It is a truly remarkable monument: an elegant but massive tower of gleaming white marble, leaning over at a quite preposterous angle and also, believe it or not, visibly banana-shaped!) The Niagara meter also works because larger particles have a larger area than smaller particles, although the meter will not measure similar volumetric flows of larger and smaller particles quite accurately, as discussed below.

THE SCIENCE AND THE MATH

The Niagara meter assumes that particles will project an area proportional to their volume on the screen. This leads to slight errors because the volume V varies as r^3, whereas the area A varies as r^2 for particles of similar shape such as spheres. Thus particles of volume $2V$ will project an area of 80 percent of $2A$, rather than area $2A$ projected by 2 particles of volume V. This is a comparatively modest error, at least for particles of not too wide a range of sizes.

The light blocked by the passing particles is, at least at low concentrations, proportional to the amount of light blocked by a single particle, multiplied by the number in the stream. A single particle is in the illuminated area for just a fraction of a second, a fraction that depends upon how fast that particle is traveling from the pouring spout. Neglect for the moment the speed of the particle just before the nozzle, and assume that the particle travels through the relatively short illuminated distance at constant speed. We have, then, for the speed v entering the illuminated area:

$$v = \sqrt{(2hg)},$$

where g is the acceleration due to gravity and h is the height down which the particles accelerate before they reach the illuminated area. The time t for which each particle is actively blocking light is

$$t = H/v = H/\sqrt{(2hg)}.$$

Now suppose that the particle is of average projected area A, and I is the intensity of light transmitted, which is I_o in the absence of particles. Then

$$I = I_o(1 - NA/[HW]),$$

where H and W are the height and width of the illuminated area and N is the number of particles in the stream within the screen view. This gives us

$$NA = HW(I_o - I)/I_o$$

which gives us

$$NV = HWd(I_o - I)/I_o,$$

where d is the average diameter of the particles.

But the volumetric flow of particles, Q, is given by

$$Q = NV/t.$$

So $Q = dHW(I_o - I)/tI_o$, giving us the Niagara equation:

$$Q = Wd(\sqrt{(2hg)})(I_o - I)/I_o.$$

This formula shows that, with the approximations we are using, the height of the screen doesn't matter but the width does. Similarly, the effect of the drop height before the screen varies only as a square root. The mean diameter of the particles enters the equation, however, and there is an intrinsic error, as discussed, because of particles of different sizes

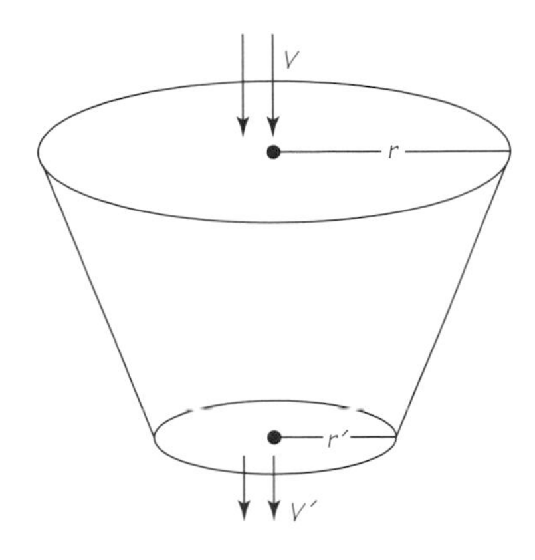

blocking light in proportion to their area rather than to their volume.

We could add many refinements to this analysis: the distortion due to some particles passing closer to the screen than others, for example, or the acceleration of the particles during their transit across the screen. We can also apply the same sort of considerations to another situation: a column of water spouting from a faucet. This should help us understand why a column of water narrows rather than thickens as it descends. The reason is that the water is accelerated by gravity. Because surface tension pulls the water together into a solid column, that column must perforce get thinner as it moves faster. You can calculate the thinning fairly easily by noting that, once the system has reached a steady state of dynamic equilibrium, the mass per second crossing any chosen cross section of the water column must be a constant:

$$v\pi r^2 = v_o \pi r_o^2, \text{ or}$$

$$v(s)\pi r(s)^2 = \text{constant} = v_o \pi r_o^2,$$

where v_o is the speed of the water stream leaving the faucet, $v(s)$ and $r(s)$ are speed and radius at a chosen distance s, and r_o is the radius of the faucet outlet. But $v = v_o + \sqrt{(2sg)}$, from the above equation $v = \sqrt{(2sg)}$, adding a starting velocity v_o as the water issues from the faucet. Hence we arrive at the faucet water-column equation:

$$r = v(v_o \pi r_o^2/(\pi[v_o + \sqrt{(2sg)}])).$$

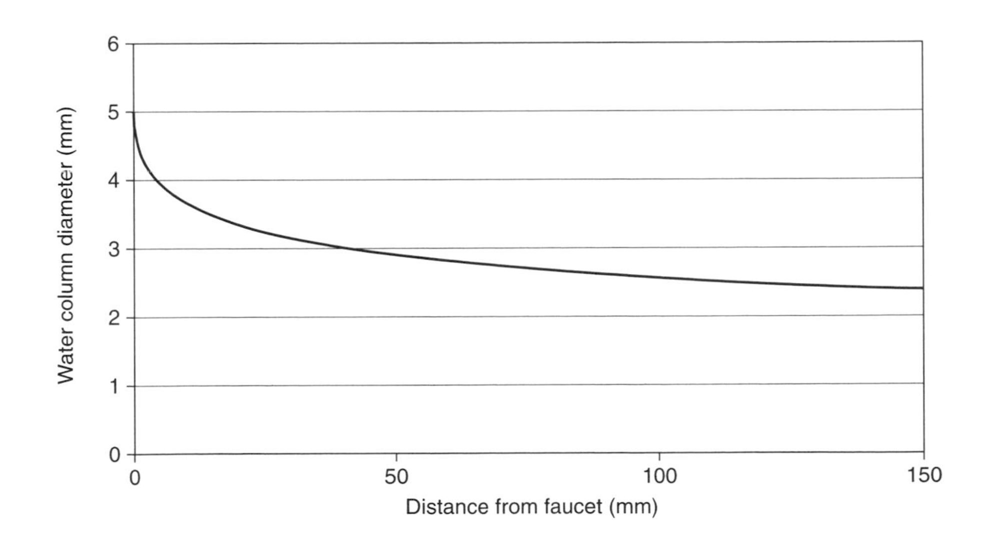

And Finally . . . Laser Optical Particle Counters

Could you use a flat laser beam instead of the illuminated screen as the light source in the meter? In industry, airborne particulates are often measured using optical particle counters. In these, particles are sucked through a capillary tube illuminated by a laser beam. The momentary scatters of light from the particles as they pass through the intense beam allow not just the counting but also the sizing of the particles, at a rate of millions per second. These laser optical particle counters (OPCs) are now part of the industrial scene, particularly in businesses that use clean rooms, such as the semiconductor industry. I once worked with a laser OPC that could also provide a measure of "spikiness" for each passing particle. This measure, which detected dangerous asbestos dust in the presence of ordinary harmless dust, relied on multiple detectors picking up the scatter of light into different planes by the laser beam.

What difference would it make to reduce the height of the screen, *H?* Is there any way you could correct for the area error in the measurement of larger particles? And what about the measurement of particles that are not round, like long rice grains, for example?

REFERENCE

Downie, Neil A. "Fibre Size Monitor." European patent no. EP0225009, 1987.

19 Magical Magnetic Valveless Valves

His most aesthetic,
Very magnetic
Fancy took this turn—
"If I can wheedle
A knife or a needle,
Why not a Silver Churn?"

. . . While this magnetic,
Peripatetic
Lover he lived to learn,
By no endeavour
Can magnet ever
Attract a Silver Churn!

—W. S. Gilbert and Arthur Sullivan, "A Magnet Hung in a Hardware Shop," from *Patience*

A permanent magnet cannot do anything with a silver milk churn, or indeed, with any other silver object. (There is an exotic exception to this rule: a powerful DC magnet field can levitate nonmagnetic objects of any sort in high pressure or liquid oxygen. This effect, known as the magneto-Archimedes effect, has been demonstrated, and there have been recent proposals to use the phenomenon to separate minerals.) If the magnet were an AC magnet, the story could be different. And if the silver were not pure metal but included nickel (and 100 percent nickel can look remarkably like silver), then even with a permanent magnet the story would be different—because nickel is almost as magnetic as iron.

The discovery that iron filings could track what happens around a magnet or electromagnet has had a profound influence on the development of electrical physics. Michael Faraday developed the use of filings to a fine art, and he pioneered the idea of "lines of force" acting through the empty space around a magnet. The force field of a magnet, which we now refer to as a magnetic field, was revealed to him and his contemporaries by the simple expedient of sprinkling filings. The whole concept of force fields in physics, a paradigm of prodigious power, probably developed its grip on the scientific minds of the Victorian age because of the striking visual effect of iron filings in a magnetic field.

The behavior of iron filings is so familiar that no one will be surprised by being able to control the flow of iron filings using magnets. You may be surprised, however, to find that particulate materials containing only a small proportion of iron filings in a large mass of nonmagnetic materials are also capable of being controlled.

What You Need

- ❏ Sand
- ❏ Iron filings
- ❏ Funnel, about 200 mm (8 inches) wide
- ❏ Magnet
- ❏ Electromagnet—buy one, or make your own using a soft iron rod and thin enamel-insulated copper wire (e.g., 28 awg, 30 swg, or 0.3-mm diameter)
- ❏ Batteries (e.g., a set of C-sized NiCd cells or other DC power)
- ❏ Wires
- ❏ Stand for funnel
- ❏ Various vessels to catch sand

Optional, for dosing apparatus

- ❏ Wide-bore tubing, about 15 mm (5/8 inch) to fit on exit of funnel

What You Do

You need to arrange the magnet or electromagnet so that it exerts a field at right angles to a narrow plastic pipe that will carry a sand mixture in a downward

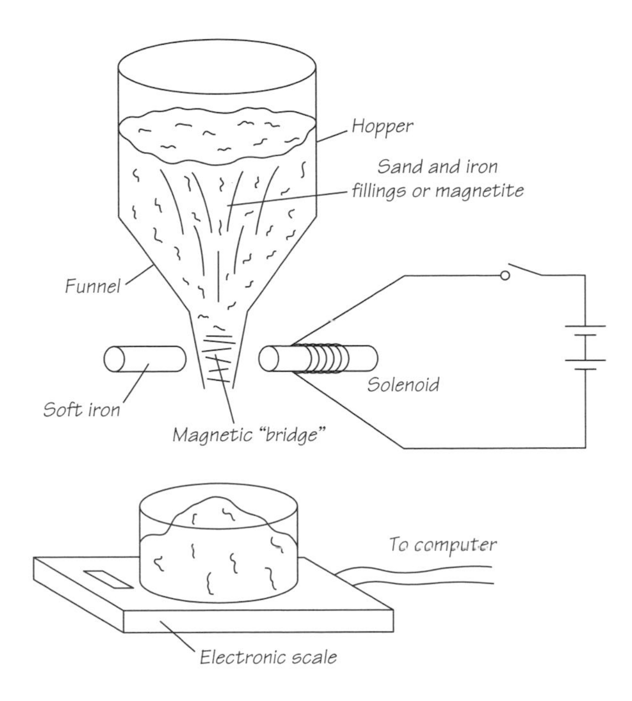

flow. I suggest you start by setting up the arrangement suggested in the diagram with a funnel and hopper, and perhaps a short length of tubing pushed over the spout of the funnel.

The sand must be fairly pure: remove other materials that might affect the flow properties, like ground-up leaves or other organic matter, tiny pieces of seashell, and so on. Sieve the sand to eliminate such contaminants.

The sand also needs to be absolutely dry. It will never be dry if it has sea salts on it—these adsorb a film of moisture—so the first thing you must do with beach sand is to rinse it with fresh water several times. If you have had to do this, or if the sand is not dry, place it on a tray in front of an electric fan heater, pick up handfuls, and let them trickle slowly through the warm air, until the sand is dry enough to flow freely. Alternatively you can dry out layers of sand 6 mm (¼ inch) thick on baking trays in a kitchen oven.

The kind of sand also matters. All sand is chemically the same: nearly 100 percent quartzite, a form of silicon dioxide. However, the sizes of individual sand particles vary from a millimeter or so down to a tenth as much or less. Even more important is the shape of the particles. Rounded particles, seen more commonly in river sand, are preferable to the typically more angular particles found in seaside sand. You can examine sand under a powerful hand lens or low-power microscope to get a feel for this—or simply let different samples of dry sand dribble through your fingers: you will easily find differences.

Try a 50/50 filings-to-sand mixture at first, measured by weight. Because the filings are two or three times more dense than the sand, this mixture will appear to be fairly full of sand. You should find that the mixture flows freely from the funnel until you apply the magnet to the lower end. Try adding more sand to the mixture and repeating the experiment, until the magnet is not strong enough to stop flow completely. I found that I could dilute the iron with sand by 4:1 by weight (between 12 and 8 to 1 by volume) and still control the flow, at least with a reasonably powerful magnet.

Try holding different permanent magnets at various points around the spout in several orientations. A north pole opposite a south pole on opposite sides of the spout should work, although there will be more and less sensitive points in the spout. Try squashing the tube a little where you apply the magnets. Experiment with a magnetic "yoke": a piece of iron that connects the pole pieces.

Once you have established a working system with the permanent magnet, set up an electromagnet in the system. Start with one C-size cell and work up to larger numbers, until you can generate a magnetic field comparable to the permanent magnet's. You should find that a surprisingly modest magnetic field is sufficient to control the particulate flow. If you can't find or buy a suitable electromagnet, then wind a lot of fine enamel-insulated copper wire around a soft iron rod. (Large old electromagnetic relays are worth taking apart for electromagnets, as are old photocopiers, which often contain solenoid coils, as noted in "Hints and Tips.") Use an electric pillar drill if you have one, rather than winding by hand—you will get better results, quicker.

Just as with permanent magnets, an iron yoke may improve performance. One way of achieving this with an electromagnet is to wind a coil directly onto a C-shaped piece of iron. For example, try grinding off the head and point of a large iron nail; then, with the aid of a hammer and vise, you can form the nail into a C shape.

Try out the hysteresis in your system: the electromagnetic power required to hold up a funnel of stationary sand is much smaller than that required to stop rapidly flowing sand. Similarly, forming a blockage when the sand is poured slowly is easier than stopping a fast flow.

With luck, you will find that you can control the sand flow with your magnetic field valve just as well as with a physical blocking valve. As far as I know, no one in industry uses such a valve, which is perhaps surprising. Our magnetic field valve has no moving parts to wear and does not, unlike many valves controlling the flow of solids, have any tendency to crush the particles whose flow it controls.

Magnetic Dosing Apparatus

By adding tubing below the funnel, plus another electromagnet, you can create a dosing system. This requires two magnetic valveless valves along the same length of vertical tube.

Once you have set up this arrangement, close the lower valve and open the top valve. Sand will flow in until the tube is full. Now open the lower valve while keeping the upper valve closed; a measured dose, defined by the length of tube between the two valves, will fall into the hopper below. As before, you should start with permanent magnets and then switch to electromagnets.

You could configure the electromagnetic dosing system with a single double-pole switch with two positions: in "hold" position, the lower magnet is on, the upper off; in "dose" position, the lower is off, the upper on. When you switch to hold, the funnel fills up the dosing tube. After you switch to dose, the fill supply is cut off and the tube full of mixture is dispensed.

THE SCIENCE AND THE MATH

The forces that the magnets exert on the iron filings are modest. How is it that they control the flow so well? One answer is that if the sand flow is not immediately stopped by the magnet, then filings will accumulate on the inside of the tube slowly. This accumulation continues until eventually the flow is stopped. The other answer is that the filings tend to form into long lines, bridging the magnets and intensifying the local field, thus providing an increased blocking effect.

Finally, sand will also spontaneously form arches, thus blocking a tube or nozzle. These arches can form momentarily and then be strengthened by the iron filings, so that a surprisingly small amount of filings will stop the flow. The proportion of iron filings that you need to add to the sand will vary both with the size of the pipe and funnel exits and with the power of the magnet.

The formation of arches also has something to do with why the proportion of iron filings needed won't vary too much with the depth of sand above the control magnet. This odd phenomenon is true more generally: the flow of particulates like sand from an orifice does not increase with the depth of sand above. The explanation is that, unlike in liquids, the formation of bridges, arches, and similar effects ensures that the flow of particulates depends only upon a limited depth of material above the orifice. But how do those arches form?

You can gain an insight into the formation of arches in particulate hoppers by considering an "arch" of just two elements in a two-dimensional hopper. The two rods are hinged in the middle, at an angle of θ to the vertical ($\theta = 90$ degrees when they are completely flat). The weight of the rods will naturally try to unfold them, increasing the angle θ. As they try to unfold, they will exert a force F_p perpendicular to the sides of the hopper, and this force will produce a friction force F_v.

This dynamic resembles the way climbers can get up vertical-sided "chimneys" on mountains. A chimney is a vertical-sided square channel up a cliff face; mountaineers push sideways with their backs on one side and their legs on the other, and they don't fall down. It is not as easy as it looks. You have to push quite powerfully against the sides of the chimney to avoid slipping, although with a chimney of the right width you can feel fairly confident, because you are wedged in and because your legs will lock in the straight position when forced downward. However, it is no good simply not falling down: you also have to be able to climb upward, which requires strength and skill.

We can also use the standard approximation for friction, which expresses the maximum vertical force $F_{v\,max}$ as equal to μF_p, with μ the coefficient of friction and F_p the force at right angles to the surface. This formula applies when the rods are just about to slip down the hopper. Taking the thrust force in the rods to be F_c, we have

$F_p = F_c \cos\theta$ (resolving the thrust in the rod to the thrust perpendicular to the wall),

and $F_v = mg \sin\theta$,

where m is the mass of the rods and g the acceleration due to gravity, resolving the thrust in the rod to the thrust down the wall. Thus we have

$F_v/F_p = \mu = 1/\tan\theta$.

Applying some numbers, we can derive the minimum angles θ that the rods must assume if they are not to slip down the hopper:

μ	θ
0.6	59 degrees
0.3	73 degrees
0.2	79 degrees
0.1	84 degrees

Thus, with very slippery walls, the rods must lie almost flat to exert enough horizontal force to grip the wall. With rough walls, however, the rods can form a tall inverted V shape and still not slip.

Now imagine that each of the rods includes a spring, representing the elasticity—the compressibility—of the rods. If you put them in the silo, they can still arch upward, and they can still be supported if they form an angle larger than the critical angle. Now, however, as the angle θ gets close to 90 degrees, the thrust in the rods compresses them. Above a certain angle θ, the rods are unstable and will simply bend downward ($\theta > 90$ degrees) until the hinged pair of rods will fall downward.

Now think about multiple pairs of these rods-with-springs lying across the two-dimensional silo. If the bottom one of the pairs is held up by a small extra force—such as the magnetic attraction of the iron filings—then the pair will be stable, even allowing for the spring's compressibility. But that means this pair of hinged rods can support a small force from a pair above, making that pair more stable. In fact, an infinite tower of rod pairs could be supported by a small force at the bottom. Conversely, if you take away that small force at the bottom by reducing the magnetic field, and the pair at the bottom becomes unstable and collapses, it won't support the arch above it; that arch won't support the arch above it; and so on. One small force at the bottom therefore controls a whole series of hinged rods with a large weight and a large net side-thrust of rods on the silo's walls. So you can understand how the "amplifying" action of the sand-jack arises.

Similar considerations also apply to a 3-element arch, as well as to a 4-, 5-, 27-, or whatever-element arch, but I won't go through the analysis

THE ANCIENT EGYPTIANS AND THE "SAND-JACK"

Curiously, the inhabitants of ancient Egypt knew how to employ the arch-forming capabilities of sand to make a peculiar kind of machine with some of the functionality of a hydraulic jack—what we might call a "sand-jack." The sand-jack comprised a piston and a cylinder, initially filled with sand and equipped with a plug valve to let the sand out at the bottom. Sand was not an obvious choice for a "working fluid." Crucially, however, the sand-jack did not require the use of precision-engineered steel pistons and cylinders, check valves, or even simple O-ring oil seals. The sand-jack could be made from strong wood by simple carpentry.

A sand-jack can be loaded with an extremely large weight, and yet the sand does not communicate much pressure to the plug at the bottom. If you remove the plug, you will see sand jetting out as the piston descends gently. Replace the plug, and the piston halts immediately. The sand-jack is thus a kind of amplifier, because a tiny force—that required to hold the plug in place against sand pressure—is magnified to the enormous forces that the piston is capable of exerting. (See the explanation in the math section.)

I have assembled demonstration sand-jacks using 75-mm (3-inch) drainpipes, with seats fitted to the pistons. With this kind of rig you could gently lower a person as heavy as 100 kg (220 pounds) down to the ground from a height of 1 or 2 m (3 or 6 feet).

The sand-jack has some significant disadvantages, the most serious being that it will only go downward. You could use it to lower your car to the ground gently after replacing a wheel, but you couldn't get the car up onto the jack in the first place.

The Egyptians used sand-jacks to complete burials in vertical shaft tombs, filled in with sand. They were often employed to lower huge stone lids—weighing many tons—onto sarcophagi. A sarcophagus lid would be placed on the jacks, perhaps by another ingenious use of sand. The shaft would be filled with fine sand and the sarcophagus lid dragged up to the top of the shaft to rest on the sand. The stone masons would then dig the sand from beneath it carefully and symmetrically. The lid would slowly sink as the sand was removed, until it rested on the four sand-jacks. The sarcophagus and the rest of the tomb would be cleaned to await the arrival of its owner, by now a mummified corpse. The mummy would be placed inside and the sarcophagus lid lowered neatly into place by pulling the plugs out from the sand-jacks.

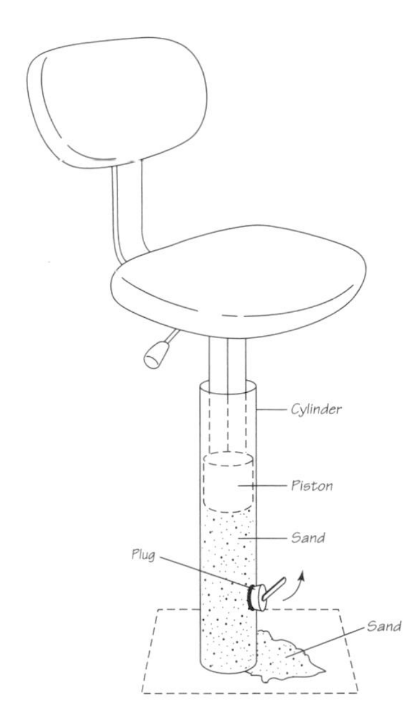

for all these! More surprisingly, the same considerations also apply to an arch that is formed from randomly placed sand grains. They too can form arches that bridge across a silo and prevent downward flow, and they too form sets of arches one above the other, with the bottom one the crucial key to supporting the whole set.

You might think that this demonstration's control of a particulate mixture flow is a matter of primarily academic interest. After all, how many practically useful mixtures include a magnetic-field-sensitive substance along with inert material? There are, however, practical systems based on this principle. For example, one researcher, Wamadeva Balachandran, at Brunel University near London, has developed a system to control the flow of grain from silos using similar effects, although using electric rather than magnetic fields.

You may also wonder why I suggest that you use such a steep-sided silo. This is to try to ensure that the apparatus will work in "mass flow" rather than "funnel flow" mode. In funnel mode, when flow is drawn from a silo, the middle of the mass of sand will sink first, forming a conical crater. As more sand is drawn out, the conical crater widens, with sand falling down the crater sides and then flowing out. The overall effect is that material from the top and center of the silo flows out first, and the bottom and edges flow out last. Particles of different sizes (such as our filings and sand) also separate during their rolling and sliding down the sides of the conical crater. There are no such effects in a mass-flow silo: instead, sand from the bottom flows out first, and there are no particular middle/edge or separation effects.

And Finally . . . Demagnetization

You could try using magnetic dopants other than iron filings. Why not try ground-up magnetite, nickel filings, or crushed ferrite? Nickel is magnetic, and it will not corrode in the presence of moisture as iron filings tend to do. The "ferrite" cores of high-frequency inductors in power supplies and other electronic devices are frequently barium hexaferrite or similar minerals. They are not electrically conductive but are highly magnetic.

The sand on many beaches includes significant amounts of magnetite (magnetic iron ore)—Fe_2O_3. If you run a magnet through such sand, you will soon pick up a considerable quantity of the ore. It would need a very powerful magnet to serve as a magnetic valve, but maybe you could devise a control or even a dosing system for this natural mixture. It is not the only such mixture, either. Magnetic iron particles are often found in china clay, for example. (China clay is the chalky white material extensively used in tableware, but also for making paper such as that which you have in front of you.)

Would it be useful to include a "demagnetization" cycle in the switch-off part of the valve operation? Many iron and steel materials that otherwise make good electromagnets retain a residual magnetic field after the coil current has been turned off, and this residual field may be enough to cause problems. (The magnetic materials used in transformers, however, generally have low residual fields.)

The application of a gradually decreasing AC current to the coil is the method of choice for demagnetization. It will cause magnetization in the iron in one direction and then in the opposite direction, canceling the previous residual magnetization. Reducing the size of the magnetization in each cycle means that, at each reversal of current, the magnetization is smaller until eventually the residual field is tiny. This sort of AC demagnetization is exactly the process followed in factories using steel parts, where there is a need to operate magnetically sensitive apparatus. AC demagnetization used to be carried out on mechanical watches that had inadvertently been magnetized by a careless owner. Magnetized watches would often stop, the mutual attractions between the delicate steel parts inside being sufficient to prevent them from working; demagnetization often restored an otherwise ruined timepiece.

What about analog control of flow? Would a continuously variable control, as opposed to the on/off control we have discussed, be possible? Would an AC magnet with a variable duty cycle be a possible way of providing that, or could a straightforward variable DC magnetic field also provide analog control?

REFERENCES

Balachandran, Wamadeva, Sidney A. Thompson, and S. Edward Law. "Electroclamping Forces for Controlling Bulk Particulate Flow." *Journal of Electrostatics* 37 (1996): 79–94.

———. "Metering of Bulk Materials with an Electrostatic Valve." *Transactions of the ASAE* 38 (1995): 1189–94.

———. "The Study of the Performance of an Electrostatic Valve Used for Bulk Transport of Particulate Materials." *IEEE Transactions on Industry Applications* 33 (1997): 871–78.

20 A Balance of Air Blasts

In cleaning mines by means of a vacuum apparatus it has been found that the ordinary implements now used, are not so suitable for cleaning the floors of mines on which the dust is mixed up with a quantity of larger material or rubble. It is found that when the ordinary vacuum implements are applied to floors covered with such rubble and dust the strong suction necessary for using vacuum cleaning in mines draws up the larger particles as well as the dust, and these tend to block up the pipes. . . . The invention consists in a suction implement comprising a slotted or perforated tube to which suction is applied surrounded by a coil guard or like device which acts as a rake and at the same time prevents the larger sized rubble or the like from entering the pipes of the suction apparatus.

—Hubert Cecil Booth, British patent no. GB190606797, 1907*

A river, although it may appear unchanged for years, is in fact a completely new stream of water in just a few minutes. This is an example of dynamic equilibrium, a concept that secretly underlies broad swathes of science. Today we are much more familiar with the concept, but only after two hundred years of the kinetic theory of atoms and molecules.

*Booth also patented cyclones for dust separation—an idea that has been rediscovered by recent vacuum-cleaner makers such as Dyson.

We now understand that something as simple and apparently static as water in a bottle is as much an illusion as an unchanging river. Inside the bottle, particularly fast molecules of water are constantly leaving the water surface and joining the air above as molecules of water vapor. In the gas space inside the bottle, the water and air molecules constantly collide and recollide. A few water molecules that move more slowly may hit the liquid water surface and be reabsorbed into the liquid. But all this fast and furious action—the molecules move at around the speed of sound—is invisible to us, because of the minute size of atoms and molecules. All we see is the average level of water in the bottle. Subtle experiments are needed to reveal the nature of dynamic equilibrium in kinetic theory.

In this project, we examine another sort of dynamic equilibrium: the rate of fall of a particle under gravity pitted against the upward drag due to the dynamic impact of a constant air blast.

What You Need

- ❑ Assorted test particles
- ❑ Rice, long grain/short grain
- ❑ Split lentils (e.g., orange) or split peas
- ❑ Popcorn (unpopped and popped)
- ❑ Tea
- ❑ Cracked wheat (bulgur)
- ❑ Any other small seeds and grains you have in your kitchen
- ❑ Vacuum cleaner (the cylinder type)
- ❑ Soda bottle
- ❑ Gauze (e.g., nylon net curtain)

What You Do

The idea behind this system is simple. Air flowing rapidly but at a controlled speed can exceed the terminal velocity of particulates falling under gravity and so carry them up. With the correct choice of airspeed, heavier, lower-drag particles (the "heavies") will be dropped and lighter, higher-drag particles (the "lights") will be lifted, thus achieving separation. A fine gauze stops the lights from escaping down the vacuum tube, retaining them as long as flow is main-

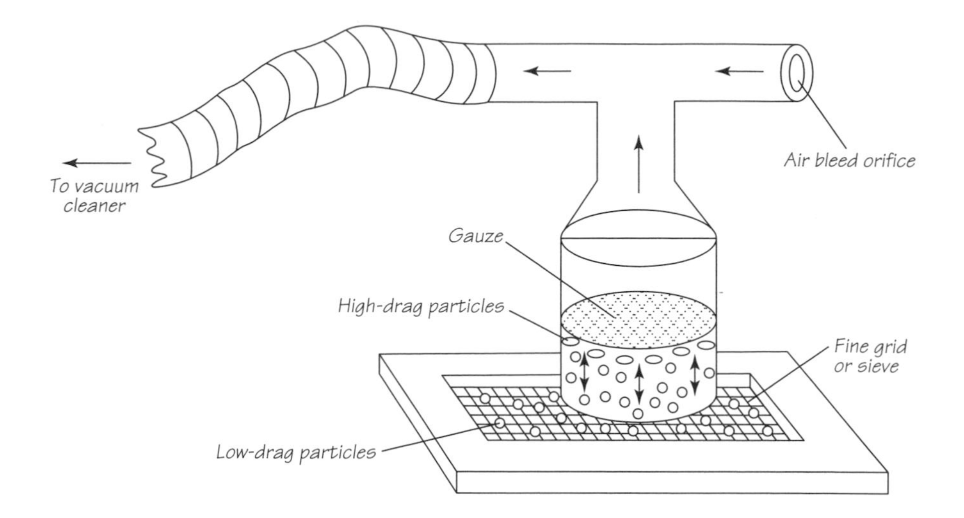

tained. You remove the lights by moving the extraction head away from the source material and then releasing the vacuum momentarily. This drops the lights into a separate bin, while the heavies remain in the source material bin.

The system behaves badly if you use the separating tube as a kind of vacuum cleaner, trying to pick up particles from a solid surface. When the tube is a long way from the surface, air flows in from the sides, and the airflow past the particles on the surface is often insufficient to pick them up. Then, in the last few millimeters before touchdown, the air blast from the side races through, and all particles on the surface, almost irrespective of weight, will be swept up. Once swept up by this high-speed blast, particulates may travel up in a confused, turbulent flow to the gauze, and some will, against expectation, be retained, both lights and heavies, reducing the separation factor.

This is why I recommend placing the mixture to be separated on a grid or sieve, with some clear space below to allow air flow; the extraction tube can then be lowered onto this sieve. With this arrangement, airflow up the tube will remain fairly well defined, roughly similar to the flow down a long tube, although with slower flow at the edges. The length of the tube allows the airflow separation to occur. If the tube is too short, many heavies momentarily disturbed may be dragged upward and accidentally retained on the gauze.

Some Surprises

Try using a mixture of lentils and rice grains. Curiously, the heavier lentils in my tests were blown onto the screen, leaving the rice behind, at least with a carefully adjusted flow rate. The explanation may be the low drag factor of round-grain rice, compared with the relatively higher drag coefficient for the half-ellipsoid lentils (they are shaped like half of an M&M candy).

Another wrinkle is that the loading of the gauze should not exceed about 70 percent. With more particulate coverage than this, airflow will be significantly reduced. This means that some lights will fail to levitate and will be left behind with the heavies. Another separation cycle will be necessary to extract these remaining lights. With a very heavy loading, the gauze will start to drop some of the particles already extracted.

THE SCIENCE AND THE MATH

This project is a type of elutriator—a device to separate or purify by washing. Elutriators take many forms. One form, for example, is an elongated water tank with an overflow at one end and a small inlet pipe at the top of the other end. Particulates suspended in a liquid enter the tank. Their trajectory depends very little upon the details of the pipe and the mixing near the inlet of the tank, and much more upon the size and density of the particulates. The trajectory can take the particles to the inlet end, the outlet end (or even the overflow, depending upon the flow rate), and all stations in between, distributing a "spectrum" of particle sizes. There are also flowing-column elutriators, which are more akin to our vacuum cleaner apparatus.

The drag D on a particle in a viscous medium is given by the Stokes equation

$$D = 6\pi\mu RU,$$

where μ is the medium viscosity, R is the particle radius, and U is its speed. Falling under gravity, this leads to a limiting downward speed under gravity, the terminal velocity $v_{terminal}$, which is given by

$$v_{terminal} = 2\rho_p R^2 g/9\mu.$$

Here, ρ_p is the effective density, (ρ_p (particle) – ρ_{medium} (medium)). This is, roughly speaking, ρ_p, since ρ_{air} is low. For tiny particles of dust, these would be the correct equations to use. However, they do not apply here; the simple Stokes equation applies only to tiny, lightweight particles in air under gravity—less than 0.001 g in mass and measuring less than 0.1 mm across, roughly.

Our particles accelerate until they reach a terminal velocity under gravity, it is true, but their drag force is given by an equation involving a semi-empirical drag coefficient C_d:

$$D = 0.5 C_d \rho_a v^2 A$$

where ρ_a is the air density, v the velocity, and A the particle area.

Hence, if $D = Mg = 0.5\rho_a C_d \cdot \pi R^2 v^2$

then $v_{terminal} = (1/R)\sqrt{(2Mg/[\pi C_d \rho_a])}$,

where R is the radius of the particle, M is its mass, and g is gravity.

For spherical solids, density ρ_p, we have

$v_{terminal} = \sqrt{([8/3][gR/C_d][\rho_p/\rho_a])}$.

This is very different from the Stokes equation for terminal speed we derived above, $v_{terminal} = 2\rho_p R^2 g/9\mu$. Our correct equation depends only slightly on radius and ρ_p, and not at all on the air viscosity μ.

Another difference that we can observe relative to the viscous fluid case is the velocity profile across the tube. Viscous flow up a tube has a parabolic velocity profile, but this is not what we see in our air-blast tube. With turbulent flow there ought to be a fairly flat profile, roughly the same velocity seen across 80 or 90 percent of the tube, with the velocity fairly constant until very close to the walls.

And Finally . . . Variations in Vertical Velocity

Could you make a continuous separation process out of this system? Could you arrange to feed in particles by a conveyor halfway up a long tube and then simply see small high-drag particles immediately moved away upward, allowing the heavier, more streamlined particles to drop out? Or could you engineer variations in vertical velocity by manipulating the height of the tubing, suspending different particles at different places in the tubing below the gauze? Or would wall effects spoil this separation?

Perceptive Illusions

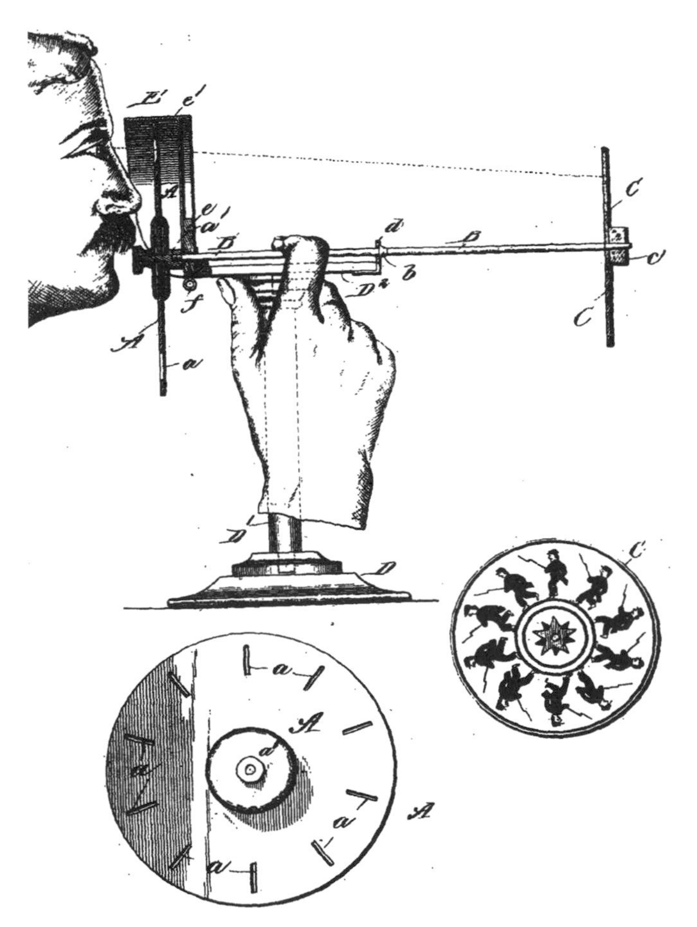

All political power is primarily an illusion. . . . Mirrors and blue smoke, beautiful blue smoke rolling over the surface of highly polished mirrors, first a thin veil of blue smoke, then a thick cloud that suddenly dissolves into wisps of blue smoke, the mirrors catching it all, bouncing it back and forth.

—Jimmy Breslin, *How the Good Guys Finally Won: Notes from an Impeachment Summer*

Most optical illusions depend upon the peculiarities of the human eye. In the hovering images project we create persistence-of-vision illusions in rotating mirrors or lenses. With the moiré microscope we produce a gridded image that is not dependent upon any particular property of the human eye, except that a certain degree of blurring of vision helps one to see it. It creates a heavily pixelated but perfectly real image in a mathematically precise way.

21 Hovering Images

Prestidigitatory illusion . . . the swiftness of the hand deceives the eye.

—Typical phrase used by masters of ceremonies in Victorian music halls

A number of toys now available—spinning tops, Frisbee-style disks for throwing and catching games—display an illuminated company logo, or someone's name, written in red light-emitting diode (LED) lamps. The ingenious aspect of these toys is that these lamps do not simply show the complete name. That would require a large number of LEDs, and the image would be blurred by the movement of the toy. Instead, there is a short row, just five or seven lamps that can, when swept through the air, create the illusion of a matrix display of many LEDs. When stationary, this row of lamps seems to glow red. They are not glowing continuously, however, but rather are switching on and off very rapidly. A simple microprocessor chip is programmed to pulse the LEDs on and off, representing successive vertical stripes of the desired display pattern. If your eye scans the row of LEDs fast enough, the words take shape. When the toy is used, its rotation ensures that the LEDs move fast enough to make the words visible.

The principle is an old one. You could argue that early televisions, based on John Logie Baird's mechanical scanner, used the same principle. In the early days of microprocessors, amateur computer nerds short of cash could display the contents of computer memories (in hexadecimal notation, to add to the magic) by using a row of LEDs mounted on a wand. You waved this lightweight stick in front of your eyes to see the hexadecimal displays of computer location contents. It was cheaper than buying a whole array of LEDs, which cost several dollars each in those days.

The basic principle is one that keeps recurring in ever more ingenious ways. A spinning top based on this principle is one of the older ideas, while the sky-writing Frisbee is one of the more recent. A still more recent development is the "air-writing clock," which displays the time hovering apparently in midair; in fact, this is a powered, decimal version of the computer nerd's hexadecimal display. Tiny LEDs mounted on the end of a thin rod are whisked too and fro through the air by a mechanism, the rod describing a sector of a circle confined between two endpoints. While it sweeps through the air, the LEDs light briefly, "painting" the time from a clock circuit in the base of the unit, while the rod remains almost invisible, its brief moment of stopping at the end of each swing hidden by the structure of the machine.

In this project we use the same principles for a new kind of display, in which the LEDs do not have to move. We also suggest the possibility of using the concept for a long-distance visual broadcasting system.

What You Need

- ❏ Pulsed LED display unit (e.g., from an illuminated-logo Frisbee)
- ❏ Motor
- ❏ NiCr resistance wire
- ❏ Circular plastic mirror (a new CD, preferably blank, will do)
- ❏ Batteries
- ❏ Wires
- ❏ Wood
- ❏ Glue

Optional

- ❏ Fresnel lens (e.g., flat-sheet magnifier)

What You Do

The arrangement depicted in the diagram is intended to move a reflected image of a row of LEDs rapidly to and fro at right angles to the row's axis. The observer will perceive this moving-line image, if the LEDs are pulsed suitably by

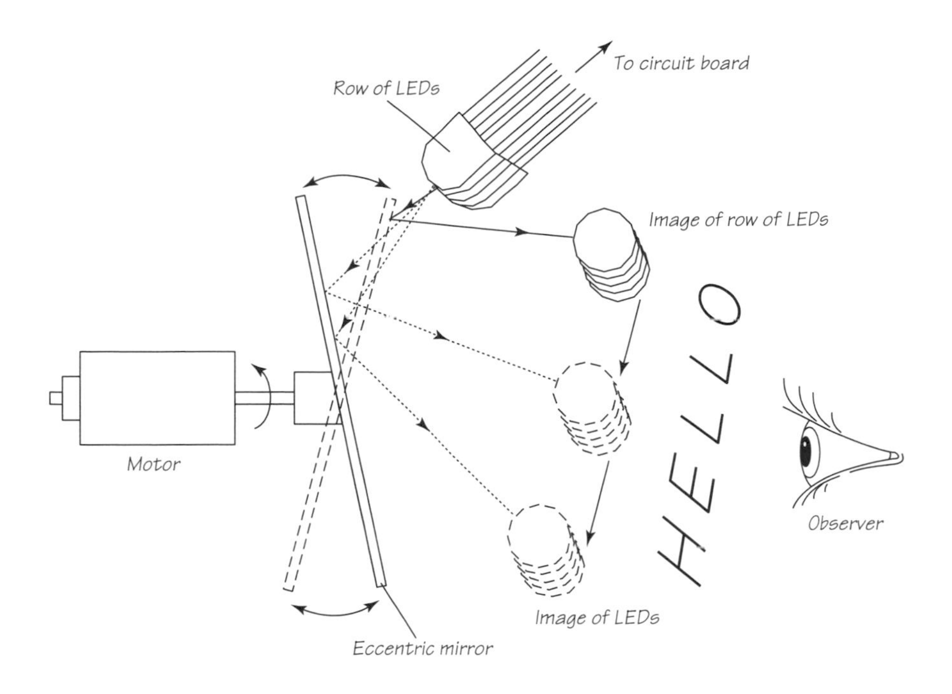

their attached microprocessor circuit, as a rectangular image of a word. You should mount the circular plastic mirror or CD on the motor so that the motor axis passes through the middle of the mirror, but with the plane of the mirror *not* at right angles to the axle. When the motor runs, you should see a reflection of the room, but blurred into circles.

Next carefully remove the illuminated logo unit from the Frisbee. Mount it as shown, so that its reflected image will be visible in the mirror as the latter is rotated from a convenient viewing angle. Most of these units can be programmed with your name, someone else's name, or other messages.

Now dim the lights in the room and start the motor, including a length of resistance wire in the circuit from the battery. Within a second or two, you should see the row of LEDs stretched out and repeated around the circle, the programmed logo becoming discernible as the motor speeds up. Try to adjust the motor speed by varying the amount of resistance wire in the circuit until the logo is readily visible.

The row of LEDs will switch on and off quite rapidly. The simple microprocessor chip pulses the LEDs, and as their light moves past your eye, your brain assembles the words out of the beams that hit your retina. We describe this process of perception as the "persistence of vision."

THE SCIENCE AND THE MATH

The eye-brain system perceives a flickering light as a steady light, provided that the flickering frequency is high enough. The critical fusion (or flicker) frequency (CFF) is around 60 Hz, a speed related to the maximum response speed of the eye's rods and cones and to the image-processing system in the retina and the brain that receives these sensor signals.

Another persistence-of-vision phenomenon is what tricks the human eye into thinking that the twenty-five images per second of movies and television are continuous, real-life action. Even before cinema, Victorian toys like the zoetrope and phenakistoscope showed how the human eye-brain system perceives a series of objects in a sequence of successive slight movements as a continuous smooth movement. The images, though, must be presented at a frequency higher than the CFF. The CFF is not a constant, however, except at extremely high overall illumination levels. The Ferry-Porter law states that, for light falling on the center (fovea) of the eye, the CFF is proportional to the logarithm of the intensity, from an ill-defined minimum of a few Hz at very low levels of illumination up to a maximum of about 55 Hz at illumination levels a million times higher:

$$\mathrm{CFF} = a \log I + b$$

where I is intensity and a and b are constants. The CFF also varies with the angle from the center of the fovea: objects at 5 degrees from the foveal axis, for example, show a CFF that is 10 Hz or so less than the CFF on axis.

For our hovering images, we don't need to suppress all sensations of flicker, but we do need to repeat the image often enough that successive images of the logo will be added up. We should also try to get as close as we can to the CFF. Fortunately, we are helped by the fact that the CFF is lower at low intensity.

We don't want the disk to rotate too fast, as otherwise the logo image will be smeared out over too long a distance. Suppose that we need the image to measure 6 mm (¼ inch) per letter, and that each letter appears for 5 milliseconds. The image must then be moving at 1.2 m s^{-1}, and, at a radius of 50 mm, the mirror must rotate at about 4 Hz, or 240 rpm: this yields a rather poor flicker rate. If we can rotate our mirror twice this fast, say at 480 rpm, then we can get up to 8 Hz, when the flicker reaches a regime where the letters are much easier to read.

And Finally . . . Broadcasting in Jigglevision

A possible advanced project would involve using a planar rotating device rather than the eccentrically angled mirror. It is possible to make a device based on an enigmatic-looking flat spinning disk, yet still see the LED image spread out in the correct way. This result can be achieved by replacing the mirror with a circle cut from the side of a Fresnel magnifier.

A Fresnel lens is one whose surface is a set of ridges. These ridges are shaped like circular glass rings that are trimmed to the same thickness and glued together

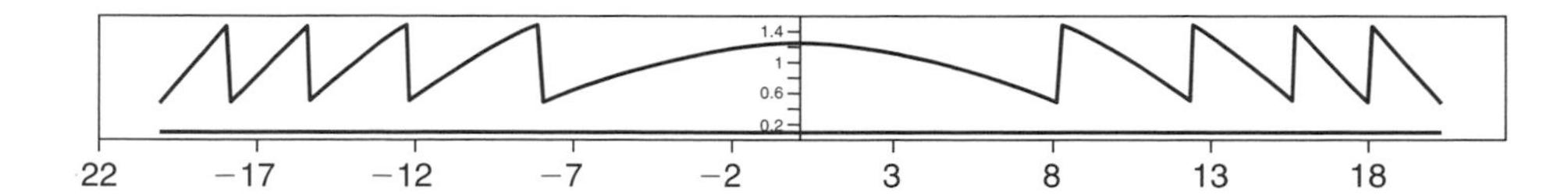

(see the diagram). The lenses—devised by Augustin Fresnel in 1822—were first used for lighthouse beams, since they offered the chance to make a lens of up to a couple of meters or more in diameter without excessive weight. These lenses are huge and remarkable objects, sometimes preserved in museums when they fall out of use. Lighthouses today still use them, as do smaller lights. I have a 1940s warship lamp—an "All Round Red"—that uses a Fresnel cylindrical lens to ensure that most of the light from the central lamp is emitted in a horizontal disk.

A small section cut from a lens, if it is not in the middle of the lens and is small enough, is simply a prism. The same is true of a Fresnel lens: a small section of it is simply a "Fresnel prism." So, in effect, your spinning disk now acts as a spinning prism. The prism deflects the incident light in different directions as it turns, and so it can spread the line of LEDs dynamically across the observer's vision, enabling the logo to become visible, just as the mirror did. You now have the option of mounting the LEDs behind the rotating circle, rather than offset in front. The effect of the Fresnel lens is to deflect the light, but it also shrinks the image, which is undesirable, although spacing the LEDs farther apart can compensate to some extent.

There is an even more advanced Saturday morning project, perhaps only suited to major broadcasting corporations: the "Jigglevision" system. Mount a set of lasers or LEDs in a vertical row on a mast 150 m (500 feet) high on top of your headquarters, each laser shining out in horizontal planes, and each being fed with a line of a video signal. Now mail out telescopes to all your viewers. The telescopes should be equipped with an electromagnetic jiggler working off the grid. The jiggler, when switched on, blurs the image sideways. Viewers only have to train their telescopes on your building's roof, switch on their jigglers, and an image will be instantly presented. Maybe a subsidiary modulation of the laser switching can carry the sound channel. With 405 lasers on your tower, 250 mm (9 inches) apart, it should be possible to present a television-like image to viewers a considerable distance away. (Can you simply remove or greatly reduce the field coil drive from a television to test the effect?)

A candle is said to be visible with the naked eye from a distance of 26 miles, presumably on a dark night. So the visibility of your lasers should not be a problem. A television picture normally would occupy an angle relative to the viewer of 15 to 30 degrees (1/4 to 1/2 radians). The 100-m-high display would occupy this solid angle to viewers out to 400 m; beyond that, a telescope would be necessary. With 50x telescopes readily available, a 20-km range would be possible, sufficient to reach to the edges of most cities.

Viewers a long way away would need more powerful telescopes. And there would be a problem or two with operating during the day. But the display would be visible at night from all positions in the city (unlike a screen) and would be relatively inexpensive (compared to a 100-m by 150-m screen with 500,000 lasers!). And without the jiggler it would not be visible, and so your broadcasting tower would not impose its image on the whole city.

REFERENCES

Gregory, Richard L. *Eye and Brain: The Psychology of Seeing.* 5th ed. Princeton, N.J.: Princeton University Press, 1997.

Sekuler, Robert, and Randolph Blake. *Perception.* 2nd ed. New York: McGraw-Hill, 1990.

22 Moiré Microscope

Where the telescope ends, the microscope begins.
Which of the two has the grander view?

—Victor Hugo, *Les Misérables*

It may seem a curious thing that a tiny hole can be used as a magnifying lens. If, like me, you suffer from a degree of farsightedness, you may have had recourse to using what might be dubbed the "squintoscope." Curl your index finger tightly into your thumb, leaving a minute hole, just capable of transmitting the barest ray of light, inside the second joint of the finger. You can then focus on bright objects just a centimeter (half an inch) or so from your eye, providing a large magnification. You can engineer the same result more precisely by making a hole in a small piece of black paper with a fine needle. This works because of the increased depth of field of your eye lens due to the tiny aperture, in effect making your eye into a pinhole camera. Here, however, we show how large arrays of small holes can be used as a sort of magnifier, using an entirely different principle from that of the squintoscope.

Moiré patterns were named after the French word for a kind of silk. Cloth woven from fine thread with a ribbed weave, so fine that it is almost transparent, demonstrates the effect beautifully, so this an apt term. Two pieces of the cloth, when overlapped, show strong wavelike variations in light and shade; this is the moiré effect of physics. Perhaps a little confusingly, moiré silk (also known as "watered silk") also refers to a technical effect used by manufacturers of cloth (not necessarily silk) in which rollers crimp the cloth to fake the real moiré effect of two layers of silk.

The moiré effect occurs because one piece of silk has a slightly different distance between its ridges than another. Suppose one piece has ridges 5 percent farther apart than the other. Put the two together with the ridges along the same axis. In one place, the ridges will line up in front of each other, letting light through and producing a light effect, while in another place, about ten ridges away, they will line up alongside and block all light, creating a dark effect. The two pieces of silk will thus show dark and light in strips ten ridges apart. Cross the two cloths over slightly, and the light and dark sections will show a more complex pattern, but according to the similar principle—light and shade in patterns whose scale is much larger than the scale of the weave. Move the two cloths relative to each other, and another surprising feature emerges: a small movement in the cloth results in a large movement in the pattern. The system is a magnifier, every bit as capable as a microscope for magnifying tiny linear movements and making them readily apparent to the naked eye.

This effect has found application in modern electronic micrometers. Minute fringe patterns like silk are drawn onto two glass or polymer rulers, and a small infrared diode illuminates a photodiode through a sandwich made from the two rulers. As the two rulers are moved relative to each other, the number of zones of dark and shade passing the photodiode are counted and displayed as the measurement. In my micrometer, the magnification by the moiré effect means that a photodiode, which is a millimeter or two across, can be used to resolve distances of 10 microns, a hundred times smaller. The same principle lies behind a much older, beloved scientific instrument, the Vernier scale. It, too, allows small distances to be measured more accurately, but it uses the human eye as its sole detector and amplifier.

What You Need

- ❏ Acetate film
- ❏ Computer
- ❏ Printer

Optional

- ❏ Xerox photocopying machine

What You Do

Assemble an array of periods on your computer screen, using your word processor or a spreadsheet program. Format them as white periods on a black background ("reverse video"). Once you are happy with the array, print it onto an acetate film page. Use a fixed-pitch typeface like Courier. Courier letters and punctuation marks all occupy the same width when printed, and we certainly want the same spacing across all the columns here. Now print out some letters

```
R R R R R R R R R R R R R R R R R R R R R R R R R R R R R R
R R R R R R R R R R R R R R R R R R R R R R R R R R R R R R
R R R R R R R R R R R R R R R R R R R R R R R R R R R R R R
R R R R R R R R R R R R R R R R R R R R R R R R R R R R R R
R R R R R R R R R R R R R R R R R R R R R R R R R R R R R R
R R R R R R R R R R R R R R R R R R R R R R R R R R R R R R
R R R R R R R R R R R R R R R R R R R R R R R R R R R R R R
R R R R R R R R R R R R R R R R R R R R R R R R R R R R R R
R R R R R R R R R R R R R R R R R R R R R R R R R R R R R R
R R R R R R R R R R R R R R R R R R R R R R R R R R R R R R
R R R R R R R R R R R R R R R R R R R R R R R R R R R R R R
R R R R R R R R R R R R R R R R R R R R R R R R R R R R R R
R R R R R R R R R R R R R R R R R R R R R R R R R R R R R R
R R R R R R R R R R R R R R R R R R R R R R R R R R R R R R
R R R R R R R R R R R R R R R R R R R R R R R R R R R R R R
R R R R R R R R R R R R R R R R R R R R R R R R R R R R R R
R R R R R R R R R R R R R R R R R R R R R R R R R R R R R R
R R R R R R R R R R R R R R R R R R R R R R R R R R R R R R
R R R R R R R R R R R R R R R R R R R R R R R R R R R R R R
R R R R R R R R R R R R R R R R R R R R R R R R R R R R R R
R R R R R R R R R R R R R R R R R R R R R R R R R R R R R R
R R R R R R R R R R R R R R R R R R R R R R R R R R R R R R
```

on your grid, but this time in black on white, with the same letter in each space, again on a page of acetate (overhead-projector) material. Give several distinctively different letters the same treatment.

If your printer won't manage acetate, then print on paper and then copy the grids onto acetate using a Xerox machine. (Don't forget to use a Xerox- or printer-compatible grade of acetate: if you use a polymer with too low a melting point or the wrong surface, you may end up with a machine jammed by molten polymer, which can be very expensive to repair.) Beware: slight size changes in the paper will be caused by heat and paper manipulation in the rollers, both in the printing and in the copying process. Be sure, therefore, to follow the same procedure for both the dot grid and the letter grid.

Now place the dot sheet above the letter sheet and manipulate the two acetates relative to each other. Start with them directly above each other, either flat on a desk with a good light on them, or held up with a bright source of light—daylight is probably best—behind. You need to angle the sheets by a few degrees, and then slide one sheet horizontally or vertically relative to the other. As you manipulate them, you should see ghostly, pixelated images of the same letter that you printed multiply on the sheet of dots, but with a different brightness from each dot. The faint, fuzzy images are not at 5 points or a similar size, however, but in any size from filling the entire page to just 10 mm (3/8 inch), depending upon the angle of the sheets to each other. If you have difficulty seeing them at first, ask a friend to hold them up a good distance away—3 or 5 m (10 or 15 feet)—and then to move the sheets slightly this way and that. You should see pixelated shadows in the shape of the letter you printed—but many times larger—pass across the sheets.

Try different sizes of dots and letters, bold or regular fonts, different grid spacings (smaller spacings are generally better), and different materials—paper versus acetate. Different illumination methods may also enhance the effect: for example, try shining through an acetate on the back with a paper letter array facing the viewer.

The slight angle between the two grids is the key to the moiré magnifier. Before thinking about the two dimensions of the moiré microscope, however, consider a related problem: "beats" between similar frequencies in one dimension. If two continuous musical notes are sounded, and their pitch or frequency is close, they are not heard as two notes. They are heard instead as one note with a tremolo effect, where the volume of the note pulses, and we say that the two notes are beating. The mathematics can be understood if you use sine wave notes with frequency f and $(f+b)$.

Note 1: $y = Y_0 \sin(2\pi ft)$

Note 2 : $y = Y_0 \sin(2\pi[f + b]t)$

where Y_0 is the amplitude of each note and t is time.

Note 1 + Note 2 = $Y_0\{\sin(2\pi ft) + \sin(2\pi[f + b]t)\}$

But $\sin A + \sin B = 2 \sin([A + B]/2)\cos([A - B]/2)$ and so, if b is very small compared to f,

Note 1 + Note 2 ~ $2Y_0 \sin(2\pi ft)\cos(\pi bt)$.

This equation represents a note of the average frequency of Note 1 and Note 2, but modulated by the slowly varying amplitude term $\cos(\pi bt)$.

Now translate this into the spatial domain. Imagine that you have a filter in which the intensity of absorption varies according to a sine wave in the x-direction: it would look like a fuzzy version of a vertical venetian blind. If you have two sine-wave modulated intensity filters of similar spatial frequency (in cycles per mm, for example), then you will get a net effect, looking through both filters, of the original filter modulated by a larger-scale sine wave. Now further translate this, still in the spatial domain, from one dimension to two dimensions, and you have the basis of the moiré microscope.

With the moiré microscope, the slight difference in spatial frequency is related to the angle between the two acetates. If the two identical acetates are first lined up and then rotated by angle α, they will exhibit a periodicity p in space as follows:

$I = I_0 \sin wx \sin py$

$I' = I_0 \sin wx' \sin py'$.

Now use the rotation matrix to obtain x' and y':

$x' = (\cos\alpha \ \sin\alpha)\ x$

$y' = (\sin\alpha \ \cos\alpha)y$

or $x' = x\cos\alpha + y\sin\alpha$

and $y' = y\cos\alpha + x\sin\alpha$.

Ignoring the sine terms for the moment, we can see that x' becomes $x\cos\alpha$. So $\sin wx'$ is replaced by $\sin(wx\cos\alpha)$, that is, the x-direction spatial frequency is reduced from w to $w\cos\alpha$. The same holds for the y-direction. The beat frequencies in each direction will be at

$B_x = w(1 - \cos\alpha)/2$

and $B_y = p(1 - \cos\alpha)/2$.

Thus we see a modulated kind of sine-wave tartan, with the modulations growing smaller as the angle α grows, and growing to infinity as the angle α shrinks toward zero.

Consider how the dots sample the input letter. In each place on the acetate, the dot has shifted by a single space relative to the letter. So each dot will display a tiny sample of the letter at the corresponding position, which is different for each dot position. The result is that the dots display a giant version of the letter with the samples spaced out by the dot spacing, rather than adjacent as they are on the letter.

We can easily model the moiré magnifier effect in one dimension with a spreadsheet. A simple one-dimensional model is not difficult, as the graph illus-

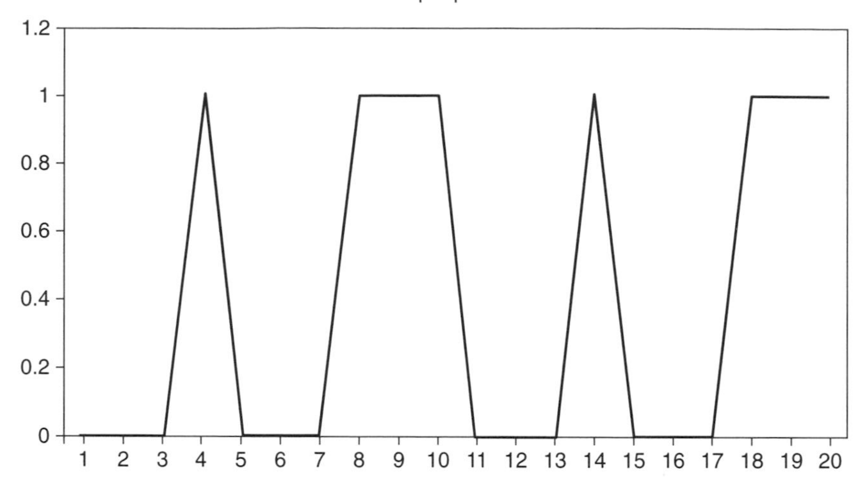

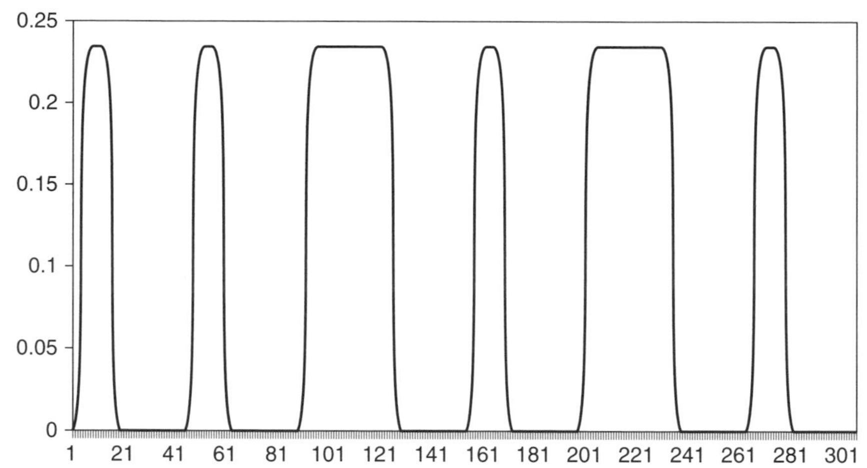

trates. In the spreadsheet, I have written a simple repeated binary sequence, repeated every ten rows. This binary sequence is then matched to a linear array of binary 1s and 0s, in which the interval between 1s is just slightly greater than the interval between repeats of the binary sequence—in this case, every eleven rows. Where the 1s match, the output column carries a 1, but not otherwise. The output column is also shown in a graphic form, with the aid of averaging and a blurring function of width for about fourteen rows. The result shown in the graph is a greatly magnified version of the original input binary sequence, with the interval between repeats now 110 apart rather than 10 apart.

And Finally . . . a Two-Dimensional Moiré Microscope

Could you produce a two-dimensional moiré simulator, perhaps with a basic computer program? The simple spreadsheet I showed for the one-dimensional case may not be enough. You may need the help of Basic or Visual Basic for the two-dimensional case.

Today, in the semiconductor industry, we have the capability to fabricate huge arrays of almost absolutely identical structures on silicon substrates: there is potential for the moiré microscope technique to be applied to the real world. As previously mentioned, it is already used by micrometers in one dimension. Perhaps someday a two-dimensional moiré microscope could be used too. Structures too small and too fragile to see directly with any kind of microscope, but which can be reproduced in thousands or millions, could be magnified and then imaged in a conventional way. The semiconductor business, or perhaps the up-and-coming industry of microelectromechanical systems (MEMS), could benefit from a 2-D moiré effect in the future.

Laser Light Shows

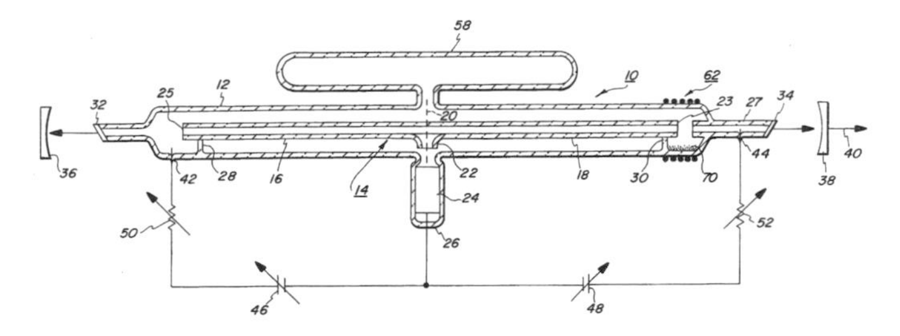

Many people said to me—partly as a joke but also as a challenge—that the laser was "a solution looking for a problem."

—Theodore Maiman, inventor of the laser, 1960*

*Quoted in Laura Garwin and Tim Lincoln, *A Century of Nature: Twenty-one Discoveries That Changed Science and the World* (Chicago: University of Chicago Press, 2003), p. 110.

Lasers have been a feature of pop concerts and outdoor firework displays for the past two decades. These events are among the few opportunities we have to see powerful lasers and the extraordinary light beam that comes from them. Lasers, though, are commonplace. The little red spot of light from a laser pointer, low-powered though it is, is still a laser beam, as is the red line that scans the bar code on items you purchase at a store. Lasers are widely used in many industries, and in most homes today there are at least three or four lasers. Typically these are small or, like the infrared light that some of them make, invisible. You don't see the lasers inside your CD or DVD players, many of which use visible laser light, and you cannot see the far-infrared light from industrial cutting lasers—although from behind a screen you can see the sparks that fly off steel when it is cut by such a laser.

Our two laser projects use two different qualities of laser light. The first, exploding laser spots, takes advantage of the almost pure monochromatic, constant wavelength of laser light. The second project, instant parabola spots, simply uses the narrowness and straightness of a laser light beam.

23 Exploding Laser Spots

> The Martians . . . are able to generate an intense heat in a chamber of practically absolute non-conductivity. This intense heat they project in a parallel beam against any object they choose, by means of a polished parabolic mirror of unknown composition, much as the parabolic mirror of a lighthouse projects a beam of light. But no one has absolutely proved these details. However it is done, it is certain that a beam of heat is the essence of the matter. Heat, and invisible, instead of visible, light. Whatever is combustible flashes into flame at its touch, lead runs like water, it softens iron, cracks and melts glass, and when it falls upon water, incontinently that explodes into steam.
>
> —H. G. Wells, *The War of the Worlds,* 1898

The absorbency of porous materials—how much they absorb, how quickly—is an important characteristic of many manufactured goods. Whenever you use a humble facial tissue, or when ink from your pen doesn't soak into the paper and spreads into an ugly blotch, you reap the fruits of centuries of research and development into the properties of paper. Porous absorbents that absorb liquids are key parts of many other devices, too. Acetylene gas, for example, is stored absorbed in a special absorbent porous plaster inside a gas cylinder; acetylene is an unstable molecule, and without these precautions it might detonate. Building materials likewise must always be studied for absorbency. Microporous thermal insulation can lower your heating costs, but how does it perform when soaked by rain?

In this experiment we use the reflected light of a laser beam to amplify in a dramatic and unusual way what happens on the surface of a piece of tissue paper or other porous material as it absorbs a drop of liquid.

What You Need

- ❑ Laser diode (laser pointer)
- ❑ Table
- ❑ Mirror
- ❑ Absorbent material
- ❑ White screen

What You Do

Depending upon the size of the batteries in your laser pointer, it may be worth attaching a pair of miniature alligator clips to the terminals inside and drawing power from larger (e.g., AA-size) batteries. Once you get the laser diode working, point it downward at about 20 degrees to reflect off a tabletop between it and a white screen. Check where the reflected spot will go with the aid of a mirror lying flat on the tabletop, so that you can easily place the light in the middle of the screen.

Take a piece of porous material—one that is not obviously a fast absorber—to use as your first sample. Ordinary cardboard or paper will work over a 10- to

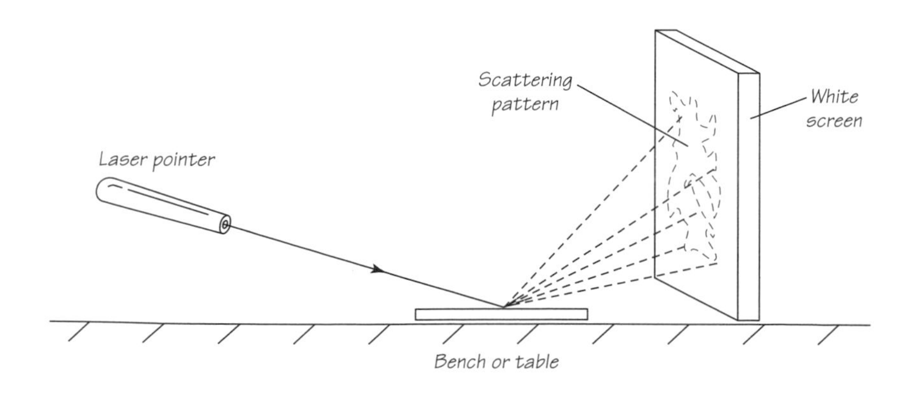

60-second time frame. Follow this test with other materials: coated papers or cloths designed to be somewhat waterproof have extended absorption periods, while highly porous materials like kitchen paper towels act very quickly.

Put each sample on the tabletop under the laser beam and turn off the lights in the room or dim them as low as possible. Because your materials are porous and have rough, matte surfaces, there should be nothing visible on the screen other than a diffuse red illumination and perhaps a little laser speckle. (Laser speckle is the curious sparkling effect—which somehow seems three-dimensional —you see when you look closely at laser light.) Now place a drop of water on the sample. You should see a clear, bright reflected spot. But wait . . . the spot will start wriggling . . . growing "veins" . . . growing larger with more wriggling veins . . . and then explode like the destruction of a Klingon battleship in a sci-fi film. Finally the pattern will dissipate outward like the expanding gas cloud of a movie's space explosion, and the screen once again will show nothing.

The liquid, at first a good mirrorlike reflector, seeps into the porous material's substrate. As the liquid surface falls, it separates into different flat zones at slightly different angles to each other. Each zone has a slight curvature and acts as a focusing mirror, so the single reflected spot splits into the multiple wriggling veins shown on the screen. Finally, as the liquid drains completely into the substrate, any liquid surface disappears entirely, leaving a matte surface once again.

As well as trying different absorbents, you can test the effects of different liquids such as alcohol or paraffin on the materials. Are they absorbed faster or slower? What about more viscous materials like oils, or mixed liquids like soap solution or sugar syrup? Do they have a similar behavior?

WARNING

As always when using a laser, avoid letting the beam, or a reflection of the beam, enter your eyes or those of anyone else. Although laser pointers are low in power—specifically to avoid eye damage—they have occasionally, when directed at a person's eyes, caused temporary blind spots or even permanent damage to sensitive retinas.

THE SCIENCE AND THE MATH

The initial spot formed on the reflector is a specular reflection like that from a mirror, slightly distorted by the not-quite-planar surface of the top of the water droplet. Once the droplet begins to be absorbed, the features of the absorbent's structure begin to show through. Some of the light spreading you see is due to diffraction effects. The angle of deflection θ due to diffraction of the light beam is the ratio of the size of the feature that causes the diffraction, S, to the wavelength λ of the light.

$$\theta \sim \lambda/S$$

gives the approximate angle in radians. The diagram shows how an interference pattern forms: this is the angle at which interference can take place and produce features in the image on the screen. Our laser light has a wavelength of about 650 nm, so when we see screen features out at 15 degrees (0.25 radians), we know we are seeing the average effects of fibers that are only about 3 microns apart.

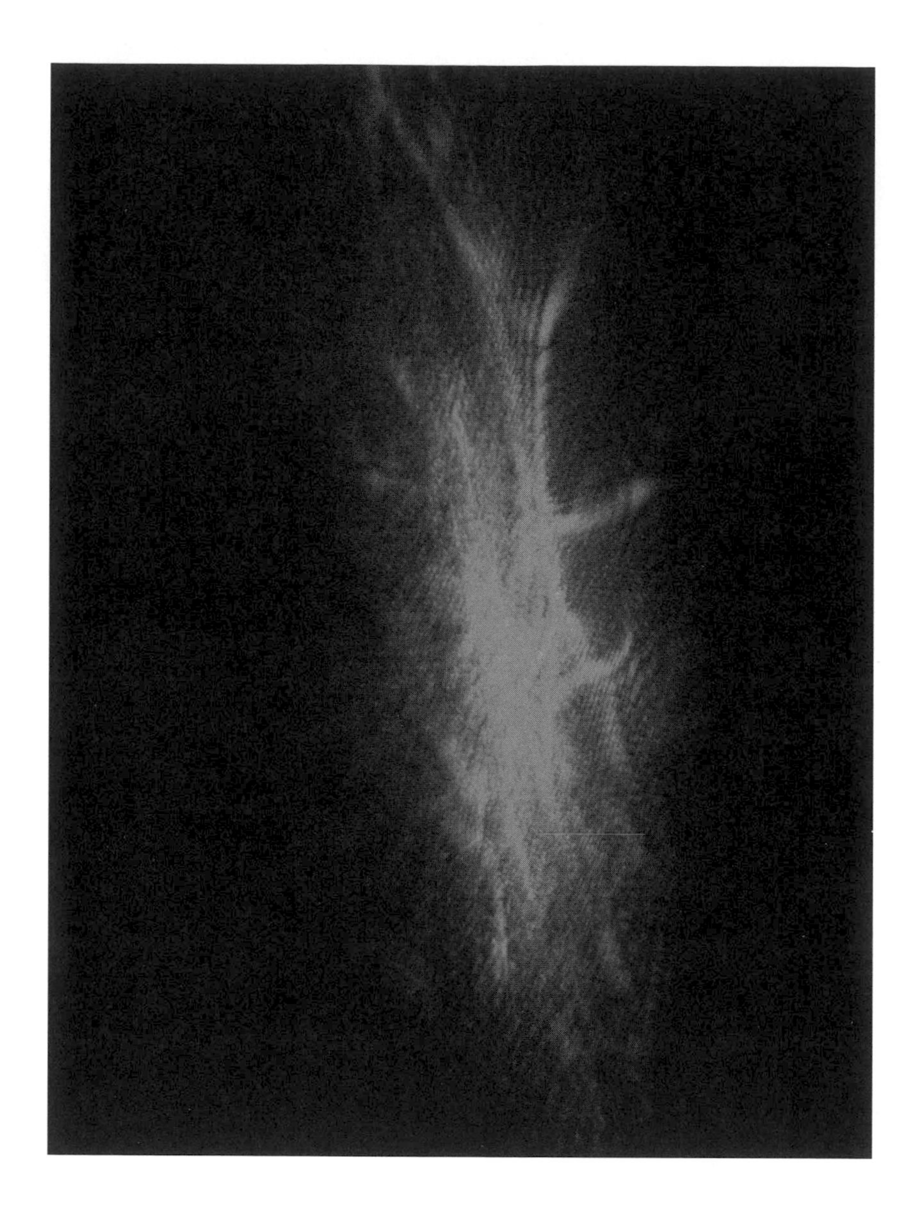

This is not the whole story. We are seeing not only diffraction but also specular reflection from tiny zones on the absorber target. When these zones are large (when the liquid is still mostly lying on the surface), they tend to be closely aligned with the surface's orientation. When they are small and can lie at larger angles to the surface, they will cause specular reflection at larger angles.

Suppose that the "roughness" of the absorber surface is caused by fibers on the order of L (say 100 microns) apart, and that the typical variation in height of the fibers is about H (say 10 microns). We should then expect that no liquid zone angle should exceed

$$\theta \sim 2H/L.$$

The factor of 2 arises because of the light-pointer magnification effect of a mirror on a light beam: when you deflect a mirror by angle θ, you deflect a light-beam incident on that mirror by 2θ. With the figures suggested,

$\theta \sim 20/100$ or 0.2 radians (12 degrees).

At first, during absorption, the pattern on the screen is created by the outermost fibers in the cardboard forming the scattering zones—small pieces of liquid surface separated by fiber dividers. The scattering reflects narrower angles because of the relatively large size of the zones and their close alignment with the surface. As the liquid drains into the cardboard, more fibers lie above the liquid level, and the zones into which the liquid is divided grow much smaller. The zones now create larger angles of light deflection.

And Finally . . . Watching Paint Dry

Different papers and cardboards produce different effects, as you have seen. But with rapidly absorbing materials—kitchen towels, table napkins, facial tissues, and the like—you need to be very quick to see the laser spot explosion. Can a digital video camera with a high speed be used to slow down the action effectively? On playback, either frame by frame or in slow motion, you should be able to see the development of the spot and time it.

Finally, what about the effect of drying oils—the resins in paints that cause them to set solid? You could devise a paint gloss development monitor, measuring the effects of the oil drying as well as its absorption by the surface. One is tempted to ask, What is more fun than watching paint dry? The retort is, Watching paint dry with a laser, of course!

24 Instant Parabola Spots

If I've told you once, I've told you a thousand times: avoid hyperbole.

—William Safire, *Fumblerules: A Lighthearted Guide to Grammar and Good Usage*

But parabolas are okay. . . . Parabolic curves pop up in many places in science and nature. When you squirt water from a hose, you get a parabola. Want to focus sunshine onto a spot with a mirror? You need a paraboloid (a parabola rotated around its axis). If you write down a run of numbers and underneath write the total of all the numbers to the left in successive places, you've created another parabola.

I was once rather intrigued by a parabola stick (I don't know what its official name is, but parabola stick sounds good to me). Tie a set of strings to a stick, spacing them evenly along the length of the stick. Attach a weight to each string at a distance that is proportional to the square of the distance along the stick. Then hold the stick out horizontally—the weights will form a parabola: no surprises there. Now angle the stick at about 45 degrees, down to 30 degrees, or at any other angle—it will still be a parabola! Use metal rings instead of weights, and double strings instead of single strings, and you can make a set of rings that will lie in the shape of a parabola, through which you can fire a projectile or a stream of water. If you get everything right, the projectile or water will shoot neatly through all the rings. But does the projectile also go through the rings when you aim the stick and projectile somewhat up? Somewhat down?

Parabolas can also pop up in the multiple reflections of a laser from not-quite-parallel mirrors. You might think that a curve like a parabola could not develop from flat mirrors and a straight laser beam—but you would be wrong.

What You Need

- ❑ A semiconductor diode laser (laser pointer) or a HeNe laser (store checkout type)
- ❑ Two large mirrors approximately 200 or 300 mm or more across, front silvered if possible
- ❑ Mountings to allow precise, nearly parallel positioning of mirrors (see diagram)

What You Do

Front-silvered mirrors are readily available from specialist suppliers like Edmund Optical. You can also sometimes find front-silvered mirrors in discarded junk—for example, inside old photocopiers. You can dissolve the plastic coating on the backs of some ordinary mirrors with a long soak in a powerful solvent like a cellulose (automotive) paint thinner; you can then use the exposed back-silvering as a front-silvered mirror. But if you can't pursue any of these options, all is not lost. Use an ordinary back-silvered mirror; the parabolic spot patterns

WARNING

As always when using a laser, avoid letting the beam, or a reflection of the beam, enter your eyes or those of anyone else. Although diode and HeNe lasers are low power—specifically to avoid eye damage—they have occasionally caused temporary blind spots or even permanent damage to the sensitive retina of the eye.

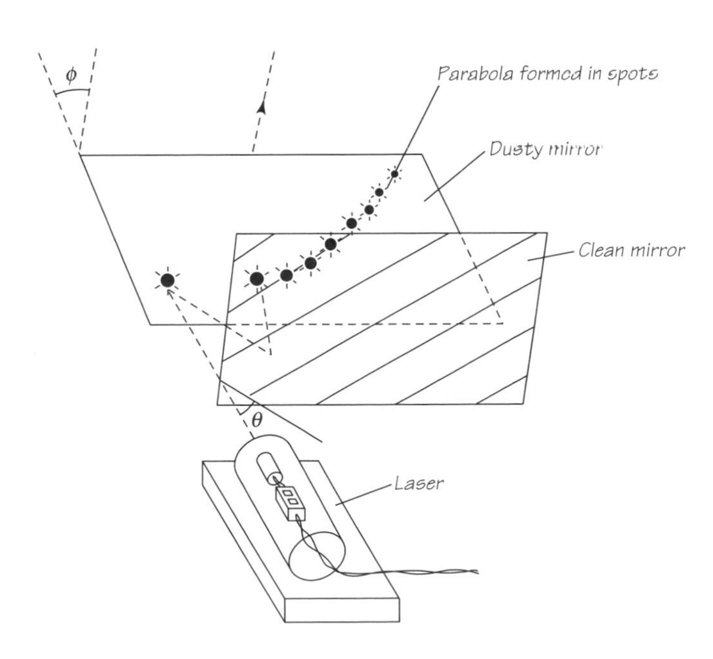

will still be there but will be more complex and interesting, and more difficult to explain.

I mounted my mirrors with the aid of identical small wooden blocks at three points around the edges. I stacked the blocks to achieve small angles and used elastic bands to hold them together. I mounted the laser on a clamp stand to allow for adjustment. The laser should be aimed at an angle nearly normal to the first mirror it hits.

With the mirrors parallel, you should be able to adjust the laser to produce ten or so successive reflections, leaving bright spots of light where it hits the imperfect mirror. The spots should lie in a straight row on each mirror. (By the way, an optically perfect mirror will not work in this project. You need a little fine dust or scratches on the mirror to scatter the light at each reflecting point, so that you can actually see the laser-illuminated spots. A little of the natural grease from your fingers—the same grease that forms the fingerprints beloved of forensic science—will also do the job.)

Now try adding a piece of packing material behind one of the blocks, and you should find that the line of spots now marches off in a curve across the mirror. Depending upon the laser and mirror angles, you can observe parabolas or straight lines. Notice that the spots are not in general evenly spaced.

You may like to use a little smoke. A small amount of smoke, such as from a cigarette lit for a few seconds and blown around the room, will produce enough minute particles in the air to render the laser beam and its many reflected beams readily visible in a small, darkened room. It is remarkable how only a little cigarette smoke is needed. You can then follow the laser beam and its many reflections, and you should be able to understand qualitatively how it ends up drawing a parabolic curve.

THE SCIENCE AND THE MATH

The parabolic plot described by the multiple reflections derives from the angle between the mirrors in the X-direction and the angle between the mirrors in the Y-direction. The angle in each direction determines the increment in the distance between spots.

With the two mirrors parallel and approximately distance S apart, the spots will follow a straight line, with the incremental changes in X and Y, ΔX, ΔY following the relation

$$(\Delta X, \Delta Y) = (\alpha S, \beta S)$$

or, as a parametric curve

$$(X_n, Y_n) = (n\alpha S, n\beta S).$$

Because S is a constant, as are α and β (the angles between the laser and the normal to the mirrors), we have equal increments in both X and Y, so the straight line has n spots evenly spaced along it.

With one mirror at angle θ to the X-direction, you will have

$$(\Delta X, \Delta Y) \sim ([\alpha + N\theta][S + \theta S], \beta[S + \theta S])$$

or, as a parametric curve

$$(X_n, Y_n) = (n[\alpha + N\theta][S + \theta S], N\beta[S + \theta S]),$$

where n is the number of bounces. This equation holds only for small angles, but it shows how the angle θ linearly boosts the distance between the spots in the X-direction with increasing number of bounces. The Y-intervals are not incrementally boosted. This leads to the parabolic form for the spots: overall the Y-spacing is even, but the X-spacing increases linearly.

And Finally . . . Curved Mirrors

With back-silvered mirrors, the whole pattern of reflections (and refractions) all becomes a little more complicated. Try back-silvered mirrors now, if you had front-silvered mirrors to start with, and try to explain the differences. There is an even more challenging analysis that you might attempt: what kind of spot patterns can you get with curved mirrors? If you can find gently curved mirrors—a large shaving mirror or cosmetic mirror, maybe—try it to see whether your predictions work out.

Of Morse and Men

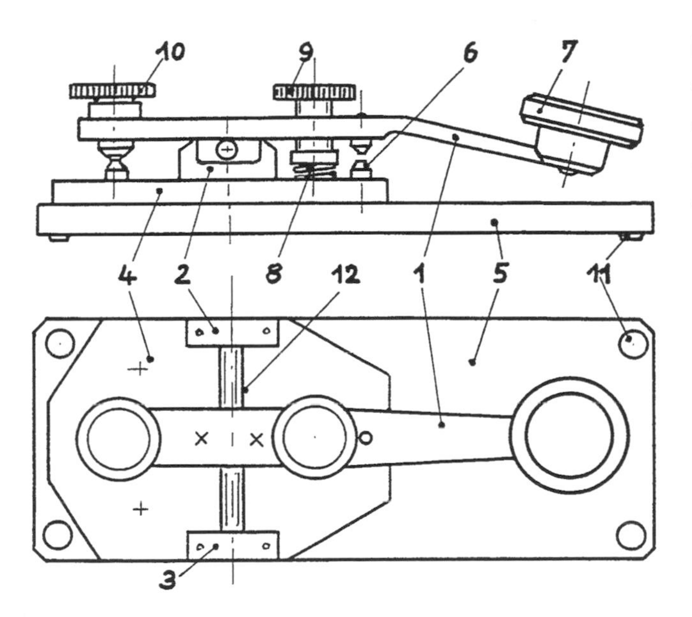

Is it a fact . . . that, by means of electricity, the world of matter has become a great nerve, vibrating thousands of miles in a breathless point of time? Rather, the round globe is a vast head, a brain, instinct with intelligence!

—Nathaniel Hawthorne,
The House of the Seven Gables

Early in the development of the science of communication, there was a need to record messages automatically at the receiving end. When pioneer Samuel Morse sent his first dot-and-dash coded messages, he wanted a system that would transmit messages efficiently and, at the receiving end, provide a faithful and unequivocal copy of what had been transmitted. His original systems included automatic transmitters, although these remained a rarity until telegraphic speeds rose above those of a human operator. Much more common and successful were Morse recorders. These provided a written record of transmitted messages, which was important for the reinterpretation of ambiguous or garbled words. In those early days, they also provided something tangible in exchange for the exorbitant fees being charged. Morse recorders quickly declined in popularity as it became obvious that, with some training, the human ear could provide real-time translation back into words and alphabet characters. Furthermore, those human ears didn't need frequent maintenance and adjustment. But although human operators were by far the most popular Morse code receivers, Morse recorders remained fairly widely in service. The same was true for later, more complex recording receivers, like the stock tickers that helped make Thomas Edison's first fortune. Many of Guglielmo Marconi's pioneer radio transmissions were received by automatic recorders.

The original Morse recorders generally used ink on paper. A common system was a lightweight inked wheel on an arm that could mark a clockwork-driven reel of paper tape. The arm was driven up and down by a relay linked to the transmission line. The wheel dipped into a small inkwell in the down position. In the up position, it was touched and rotated by the passing strip of paper. After a few preliminary dots and dashes, the whole wheel would be inked and would correctly record on the paper tape all the signals it had received. The operator could then glue the strips onto sheets of paper in rows, with the characters handwritten below, to form the message. With a weak radio signal or poor signal conditions—perhaps caused by thunderstorms—on a land transmission line, the tape could require some interpretation. And, of course, if a legal case rested on the words communicated, the primary record of the dots and dashes was much better evidence than a handwritten message taken from audible Morse.

Our first project constitutes a halfway house between handwritten Morse and automatically recorded signals. The device requires a human operator but needs no special training, simply a certain knack for holding a pen correctly. It could have been used in 1836, when Morse code systems were invented, although I have no evidence that anything like it has ever been used.

The second project explores another sort of recorded-message telegraphy system: the fax machine. A Scottish farmer and clockmaker, Alexander Bain, invented the concept of the fax machine shortly after Morse code, in 1843, but the fax was not developed for some years, and then only fitfully—you can read all about this in Solymar's *Getting the Message*. We approximate a kind of fax machine by using a toy writing tablet and a curious spiral scanning technique. Finally, we bring our communication projects right up to date with a system I call the calculator communicator, which demonstrates on a paper calculator roll the principle used by all modern communications, binary digital codes, and records transmissions.

25 Messages in Wriggling Lines

CQD POSITION 42.26N 50.14W REQUIRE ASSISTANCE STRUCK ICEBERG

```
-·-·  --·-  -·· /
·--·  ---  ···  ··  -  ··  ---  -·/
····-  ··---  ·----  -····  -· /
·····  -----  ·----  ····-  ·-- /
·-·  ·  --·-  ··-  ··  ·-·  · /
·-  ···  ···  ··  ···  -  ·-  -·  -·-· · /
···  -  ·-·  ··-  -·-·  -·- /
··  -·-·  ·  -···  ·  ·-·  --·/
```

—Message sent by Jack Phillips and Harold Bride, wireless telegraphy operators onboard the SS *Titanic,* April 15, 1912

The pioneering period of current electricity was in the 1800s. Although electricity generated by static machines was well known before then, static electrical machines were rarely capable of more than nanoamps or, at best, microamps of current, and thus could operate only at high voltage. At high voltages, the materials readily available before 1800 (such as wood, brick, or string) were poor insulators, especially in damp weather, and the resulting hazards limited the applications of electricity. The availability of currents on the order of milliamps or more from chemical batteries such as the Daniel cell, based on copper and zinc, changed all this. So-called current electricity needed only a few volts to transmit appreciable power along relatively crude metal wires, and such low voltages could be insulated using ordinary materials. The new electricity could be easily generated and easily transmitted, opening up countless new applications.

There was, however, still a problem: the detection of electricity was not easily accomplished. For example, it was known from the earliest work that a tiny wire can be heated by an electrical current and thus seen glowing. But the wire must be hairlike in size if it is not to require many amps to heat it. The hotness of the wire is difficult to see if the temperature is low, while a high temperature may oxidize the wire and destroy it. These considerations formed the basis for the lightbulb, but that was many decades later, with the application of much more sophisticated science. A more sensitive early method used electrical current to generate bubbles in salt water by electrolysis. But with a current of less than an amp or two, the tiny bubbles formed very slowly and were difficult to see.

By contrast, the detection of even microamps of electricity at a few volts is possible if electricity is fed into an animal muscle. This was an early observation; indeed, the discovery of current electricity and chemical batteries was interlinked with animal muscle contraction, as recorded by Luigi Galvani. It was, therefore, natural to try to use muscle contraction to detect current in a telegraph. Frogs' legs and the like, however, proved to be notoriously short-lived and unpredictable materials. What better, then, than to simplify and reduce costs by adding a human receiver and adding that human receiver in a direct way—as a circuit element in the electrical detector of transmitted pulses? Some early telegraph systems thus included a "human element" in quite a literal sense!*

In the early days of Morse code telegraphy, it was thought desirable that transmissions should be recorded, typically on long paper strips. Few details survive, but presumably, pioneer telegraphists would have tried to include a human element in their Morse recorders as well as their simple receivers. The user probably placed his hand in the circuit, allowing the transmitted current to involuntarily contract his hand muscle and push the inker down on a moving paper tape, which presumably he could wind off from a roll of tape with his other hand, thus minimizing the mechanism involved.

The discovery of the magnetic effects of current by Hans Christian Oersted and Michael Faraday paved the way for more efficient detectors, which in time moved the human telegraph receiver back into the cupboard of laboratory curiosities from whence it had come. But maybe our Morse pen project sheds some light on what might have been.

*I have tried this myself, maintaining as safe a system as possible, by using a Morse key on a transcutaneous electrical nerve stimulus (TENS) machine, which is an inexpensive ($50–100) medical device that delivers tiny electric shocks safely through skin electrodes as a means of pain relief. TENS machines are driven by a small battery and are guaranteed to be harmless, although they are uncomfortable if incorrectly adjusted. I set my TENS machine to deliver a continuous sequence of small pulses to my finger: the result was a tingling sensation that quite clearly indicated when the Morse key was pressed on the other side of the conference room by an assistant from the audience.

What You Need

- ❑ Pen, ideally a narrow marker pen
- ❑ Small electric motor
- ❑ Eccentric weight to mount on motor

- ❑ Powerful battery (e.g., 6-V sealed lead acid or larger capacity NiCd/NiMH battery)
- ❑ Wires
- ❑ Long sheets or continuous strips of paper
- ❑ Morse key or other similar push-to-make switch
- ❑ Glue

What You Do

This project is simplicity itself. Wire up the parts as shown, and then ask someone to transmit while another person tries to draw simple straight lines across a piece of paper. If you can try drawing along a continuous roll of paper, then so much the better.

When the transmitter provides current to the motor, the motor rotates rapidly and, because of its eccentric weight, the pen wiggles from side to side. The pen thus produces a much thicker, slightly scribbled line, instead of the usual single thin line drawn when the motor is stationary. A "Morse code pattern" can be easily perceived in the wriggling lines.

When the motor rotates, the pen bucks as though something is alive in your hand. The feeling of something else taking over your body is a relatively unusual sensation. Cramps taking over a muscle or two in your legs is an unpleasant example; more pleasantly, riding a tandem bicycle feels odd to most people, at least when they ride on the backseat. Holding a small animal like a rabbit that, after sitting calmly in your arms, suddenly panics and starts to struggle, has something of that surprising feeling too. With the Morse pen, you are happily

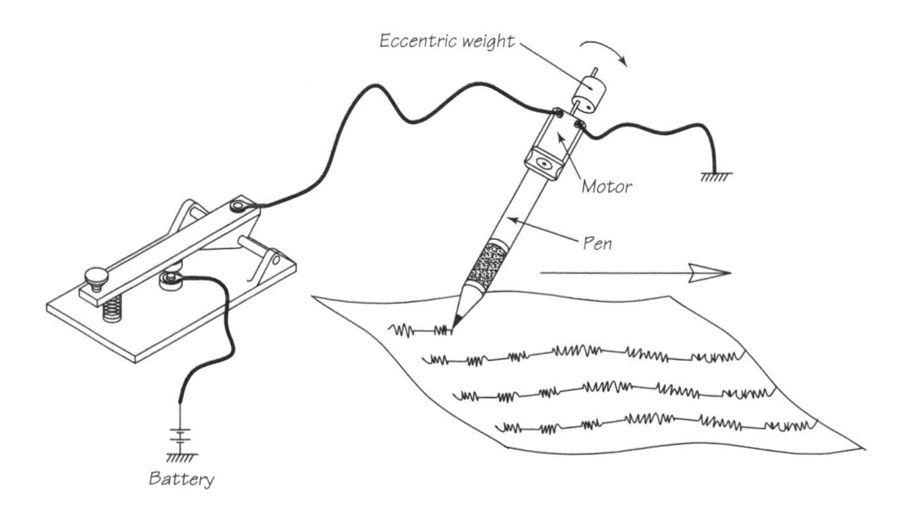

drawing a nice straight line, when suddenly the pen does its own thing, scribbling over the paper.

If the transmitting person at first produces a continuous, predictable pattern, this will allow the receiving person to practice the right sort of grip for obtaining the best results. If you grip the pen too hard, the code pulses will be difficult to discern. If you grip too loosely, the pen will wander and skip, and it will also be difficult to keep up a steady straight line between pulses.

I have used whole sets of Morse pens in lecture demonstrations, with ten people struggling valiantly to draw straight lines while the code comes over to them. Usually two or three people will succeed in getting the hang of it in the two minutes allowed, but the project will work better with a little practice.

THE SCIENCE AND THE MATH

It is difficult to estimate exactly what the pen tip will do on the paper, owing to uncertainty over how hard the pen will be pressed to the paper and how firmly the pen will be held. In the absence of a strong grip, the pen tries to keep its center of mass in one place. This will be increasingly the case the faster the motor runs. If we take the off-center mass as m, at a distance of r from the axis, and the motor mass as M, then the motor will orbit at radius R_m, such that

$$R_m \sim rm/M.$$

If the pen is of a typical light weight, then we can, to first order, neglect its mass. We can then model the grip of the person operating the Morse pen as a simple pivot distance L_g from the writing tip, so that the width W of the mark made by the pen when the motor is rotating is given by

$$W = 2R_m L_g/(L - L_g).$$

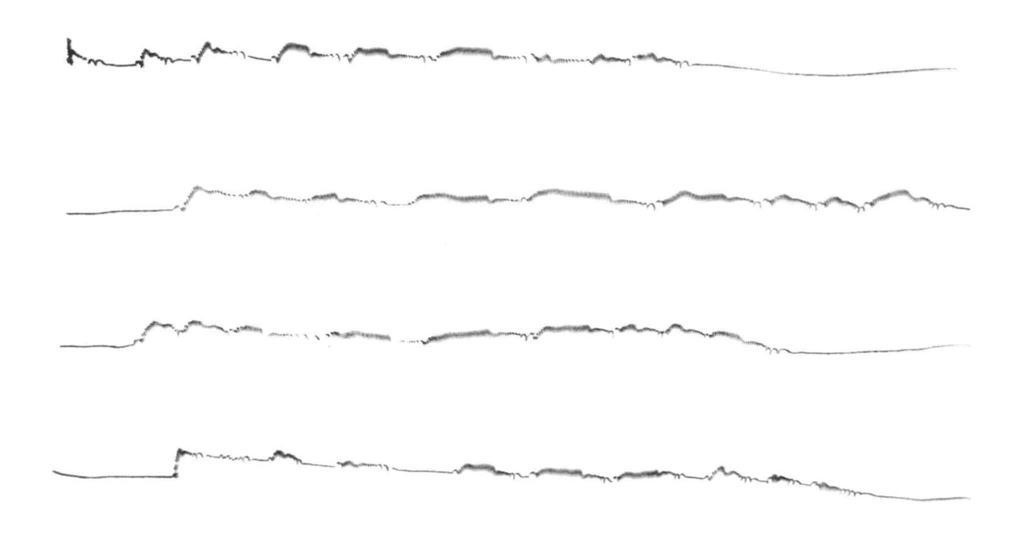

Hence, $W = 2rm/M(L - L_g)$,

where L is the pen length. To maximize the mark width W, we must maximize m and minimize M and the distance to the operator's grip from the motor, $L - L_g$.

What are the effects of a tighter grip by the operator, however? Or a more massive pen? The more massive pen can be dealt with by adding an additional effective mass M_{peff} to the motor mass M. The effect of an operator's tighter grip could also be dealt with in the same way, by supposing that the tighter the grip, the more the operator's hand will be moved by the pen's wriggling, and the higher the effective mass of the pen. These variables add another effective mass M_{heff} to the motor mass:

$W = 2rm/([M + M_{peff} + M_{heff}][L - L_g])$.

And Finally . . . Chartagrams

Morse pens are not the only way to transmit and then encode information in a wiggling line. I've mentioned, in a note, that you could engineer a different way of signally directly to a person's hand via tiny electric shocks. But are there other simple mechanical solutions that can respond rapidly enough to produce a faithful record of Morse code? With chartagrams (see the next section), an analog signal can be used to directly send letters in an unusual way.

ANALOG MESSAGES IN WRIGGLING LINES

In the early days of Morse code, complicated assemblies of relays, wheels, clockwork, and ink rollers were beautifully fashioned from polished brass and varnished mahogany. These complicated devices, however, merely recorded the dots and dashes of Morse code. There seems to have been no early attempt to write letters directly until the advent of "stock tickers" decades later. These devices wrote a continuous line of letters by scanning what was, in effect, a row of six- or seven-dot Morse inkers along a tape.

Why didn't the nineteenth-century pioneers try using an analog signal, rather than an on-off signal, and try direct writing of the letters in some way? Perhaps there were attempts to use an analog signal in conjunction with something like a chart recorder—what I call a chartagram—but I have not discovered any evidence. I think that the Victorians *should* have invented the chartagram, but they didn't, and so I have been forced to invent it for them.

My chartagram uses a single line, wiggling up and down to simulate the mountains and troughs of the top of a slightly modified alphabet. You transmit the line by a continuously varying voltage signal, one that can take any value between zero and V_{max}, and that changes with time. At the receiving end, it is plotted on paper by a chart recorder (although you could use a data logger and computer, as are usual today).

The voltage signal is produced by scanning along a "silhouette" of alphabet letters with a solar cell, modified by the addition of two pieces of tape that

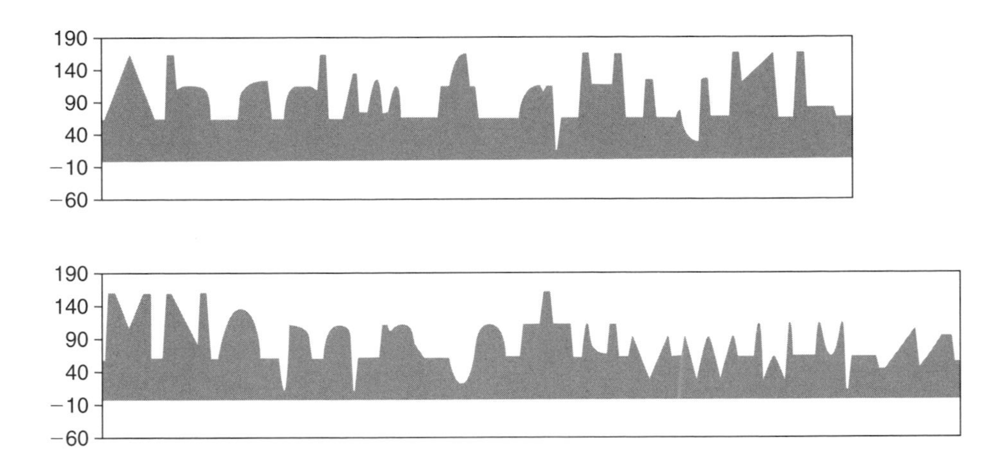

mask off all but a narrow slit down the middle. The cell produces a current proportional to the length of the slit exposed and thus to the height of the letter silhouettes. (A similar mechanism was used in film production until quite recently: a dark band of varying height encoded the sound track along the edge of the strip of plastic on which the visual frames were printed. The solar cell in this case was connected to an audio amplifier rather than a chart recorder.)

Because the chart recorder only runs forward, the wriggling line cannot draw any "re-entrant" features in the letters it draws, nor can it (unless another control channel is provided) lift the pen off the paper. However, with the aid of a special alphabet, such as the one illustrated, it is nevertheless possible to represent all the letters of the alphabet distinctly and recognizably. Using a simple stencil cut from a plastic sheet with a sharp knife, plus a black marker pen, you can fairly quickly compose messages for transmission. I found that kids could produce some recognizable silhouettes in this way during the course of a brief lecture demonstration. I don't think you need a scanning mechanism—just a steady hand to draw the solar cell across the silhouette smoothly.

The chartagram is subject, like any electronic system, to noise. The higher the impedance of the transmission system, the more noise will affect the smoothness of the chart recorder output, since a high impedance system releases less energy into the system or, to view it another way, needs less noise energy to produce a given voltage. If you try the chartagram system over any but the shortest of wires, you should use a fairly low impedance—which a solar cell indeed provides. If you need to demonstrate the chartagram over a long distance—of more than 50 or 100 m—you could adopt the 4- to 20-mA current transmission system, rather than relying on a voltage system.

The 4- to 20-mA analog telegraphy system transmits analog signals around chemical and other processing plants. It has a number of attractive features even today, when we generally prefer to send digital rather than analog signals. First, it operates at a maximum of 20 or 30 volts, so it is safe for people to touch the bare wires connected to it. Second, it has an offset zero: if a wire is disconnected, a fault is immediately obvious and can be used as an alarm signal. Third, and perhaps most subtly, it overcomes problems with resistance in wires and imperfect connectors. Any increased resistance in the circuit is ignored, since the transmitting unit detects any change in resistance in the circuit loop and increases the voltage it applies until the current once again reaches the correct value.

Could the kind of letter forms shown in a chartagram be more widely useful? When we learn to write at school, maybe we should be taught a chartagram alphabet rather than the various sort of "copperplate" letter forms that most of us struggle to write tidily. Maybe a special chartagram alphabet could also replace the squiggles and loops of secretarial shorthand.

26 Helidoodle

> "I have a mysterious letter written in a language quite unknown to me and delivered by some magical agency right on to my bed." [said Colonel Dedshott of the Catapult Cavaliers] "Most strange," he [Professor Branestawm] said, wriggling the paper about and smelling it and tasting it and listening to it. "It isn't any language I know, and I didn't think there were any others. Can't be Japanese, because that goes up and down and this goes sideways. It isn't idiomatic Crashbanian, because that's written round the edges of the paper."
>
> —Norman Hunter, *The Incredible Adventures of Professor Branestawm,* 1933

There is a popular children's drawing-screen toy called the Magnadoodle. You write on the white screen with a tiny magnet, which leaves a gray-black trail wherever you touch the screen. When you are finished, you can erase the drawing completely by sweeping a large bar magnet across the back. It is a modern equivalent of ancient Egyptian wax tablets, which were also used by children.

The whole future of imaging—not just for kids—may include an electronic system based on a technology similar to the Magnadoodle's. The "e-paper" product of E Ink Corporation is an electrostatic analog of the Magnadoodle. In e-paper, electrostatically charged particles travel through an opaque medium, under the influence of electric fields imposed by the electronic driver circuits on the system of X and Y electrodes. In this way a pixelated screen image can be produced, just like a liquid-crystal computer screen, but with an important difference—the image remains after the power is switched off.

Here we demonstrate the principle of an unusual kind of fax machine, scanning in a transmitted image (but using a spiral rather than the traditional square raster) and then reproducing it at a remote receiver on an ecologically sound reusable media: the Magnadoodle screen.

What You Need

- ❑ Magnadoodle screen
- ❑ 2 old phonograph turntables (for vinyl LPs)
- ❑ Light wooden rods or Erector-set strips for the two turntable arms (or adapt the fitted arms)
- ❑ Large nail or other steel rod at least 50 mm (2 inches) long and 6–9 mm (1/4–3/8 inches) in diameter
- ❑ Thin enameled wire (transformer winding wire)
- ❑ Thread
- ❑ 2 electric motor and reduction gears, or string and pulleys
- ❑ Photodiode
- ❑ Transistor with small signal (e.g., BC184, BC107, or the like)
- ❑ Power transistor (e.g., MJE15028 or any other with a power handling of 10W or more and a gain [Hfe] of 40 or more)
- ❑ High brightness red LED (if used, see below)
- ❑ 6-V battery or power supply
- ❑ Resistors (see circuit diagram)
- ❑ Wood
- ❑ Glue

What You Do

The assembly shown is designed to scan the input in a spiral raster: as the input picture and its companion Magnadoodle screen rotate on the turntables, the arm bearing the photodiode causes the transistor to switch on and off. This switching makes the electromagnet mimic the photodiode and hence reproduce the transmitted picture.

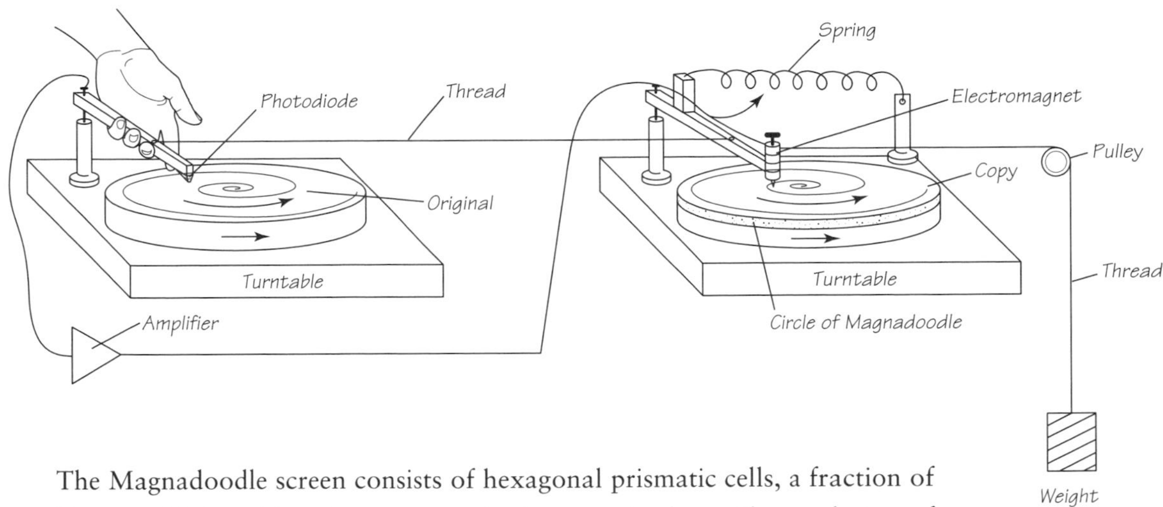

The Magnadoodle screen consists of hexagonal prismatic cells, a fraction of a millimeter thick and just 4 mm across in the screen's plane. The resolution of the system is better than that of the cells, with the smallest object displayable only about 2 mm by 2 mm. Inside are specially prepared iron filings and a liquid containing opaque white powder. Normally the filings lie at the bottom of the cell, hidden by the white powder, and the cell looks white. When a magnetic field is applied to the surface, the filings are drawn up, squeezing the opaque white powder down and turning the cell black. (The back surface image is a negative of the front surface—black, with a white area where the magnet has drawn up the filings.)

Trim the Magnadoodle screen to form a circle, sealing the edges from leaking the white liquid and filings using clear sticky tape. Place the screen on the receiver turntable, which you have equipped with an arm carrying an electromagnet. You can make a simple electromagnet with thin transformer wire wound around half of a large, 100-mm (4-inch) nail. Wind enough so that the coil has a resistance of about 7 to 10 ohms. File the end of the nail so that it is flat and about 2 or 3 mm ($^1/_{16}$ to $^1/_8$ inch) across.

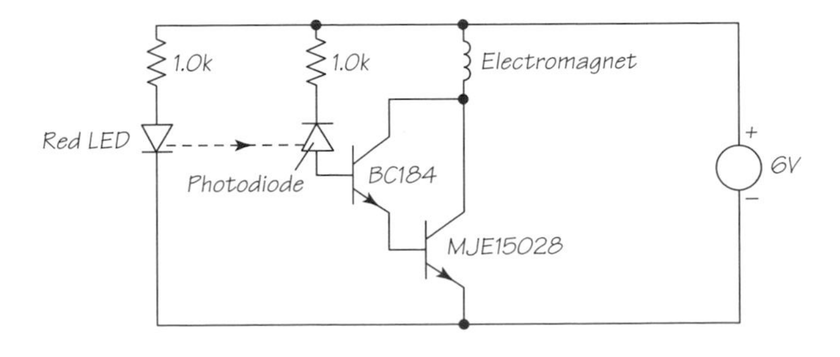

You will need to make two arms: one on the transmitter for carrying the photodiode (and LED if used), and one on the receiver for carrying the solenoid. You can use the original tone arms, removing the phonograph cartridge and needle, or you can make something similar yourself: simply a piece of wood pivoted to swing over the disk area.

Equip the transmitter turntable with a photodiode/LED on its arm, which can scan across the image you want to transmit. Ensure that the LED points at 45 degrees to horizontal toward the photodiode (also oriented at 45 degrees to horizontal but facing the LED). Alternatively, you can adopt a system that does not need light from an LED but uses ambient light. Arrange the photodiode to lie gently on the paper surface, sensitive face down. Ambient light from around the photodiode will activate the circuit when there is reflective white behind it; with black behind it, the ambient light is blocked and the circuit will not activate the electromagnet.

Trim a sheet of white paper to a circle, and then write your original message on it boldly with a broad, black marker pen.

The rotation of the two turntables is powered by their motors. But without the spiral tracking of an LP, there is no automatic radial scan. Thus, the scan should be motorized with identical motor/gearbox units to power the two phonograph arms, or it can be achieved more simply by connecting the two arms using string and pulleys, as in the diagram, and pulling the transmitting arm over slowly by hand. In the motorized version, check that the two motors do indeed scan across their disks in about the same time.

How It Works

Try the system out, scanning your input paper disk and starting the receiver scan simultaneously. The resulting image on the Magnadoodle will be a negative on the front face, with a positive on the back face. Adjust the scanning speed so that the spiral scan will match the width of the receiver line. This factor will be largely determined by the size of nail you used to wind the receiver coil. You should aim to achieve just a little overlap between successive lines in the spiral.

It is easy to make a permanent record of a transmitted image: you can copy the Magnadoodle screen onto paper with a photocopier. You may need to play with the settings on the machine, but the image should photocopy recognizably, even if the hexagonal grid also tends to be printed to some extent. With a computer scanner you can probably do even better.

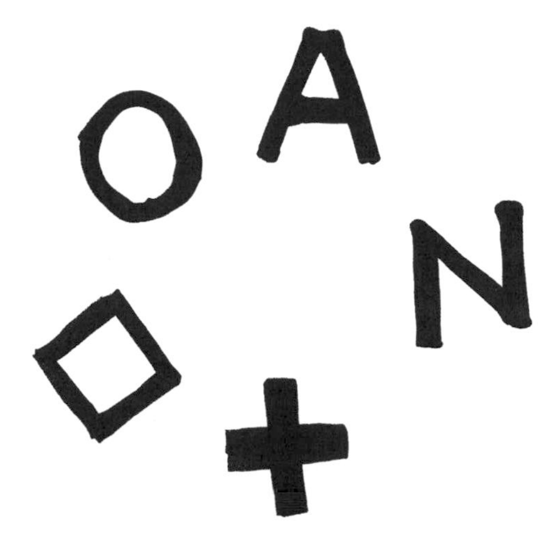

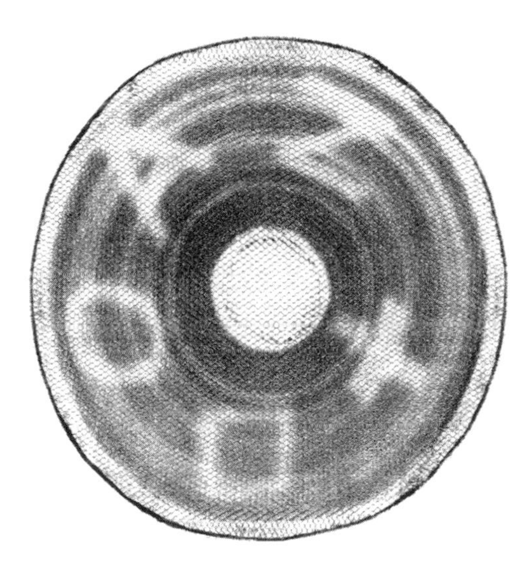

An image can be understood as a variation of intensity (or color) across a two-dimensional, usually flat surface. A communication channel, by contrast, is a one-dimensional variation of intensity with time. To convert a two-dimensional image into a one-dimensional signal thus requires a scanning method.

There are number of possible ways to scan a surface. The standard method is a Cartesian or raster scan, in which the readout point—the point whose intensity becomes the signal—is scanned fast along a line in the image's x direction, this line itself moving slowly down in the y direction. This is the basic system used in television systems, which is usually achieved by means of fast electronic circuits.

A more exotic method uses a plan position indicator (PPI) scan: with this technology, the readout point is scanned quickly along a radius, while this radial line itself slowly rotates around a central point. This system is familiar from radar sets used in aircraft by the military and in air traffic control.

Once a two-dimensional image has been scanned to form a one-dimensional signal, and that signal has been transmitted and then received, the signal must be converted by an inverse scanning method back to its two-dimensional image form. In television receivers, this result is readily achieved by electronic circuits, which scan a beam of electrons across a phosphor screen in a Cartesian raster scan matching that of the transmitter. There are other ways, however. The Nipkov disk, for example, was used in the earliest television receivers devised by John Logie Baird and his associates. A Nipkov disk

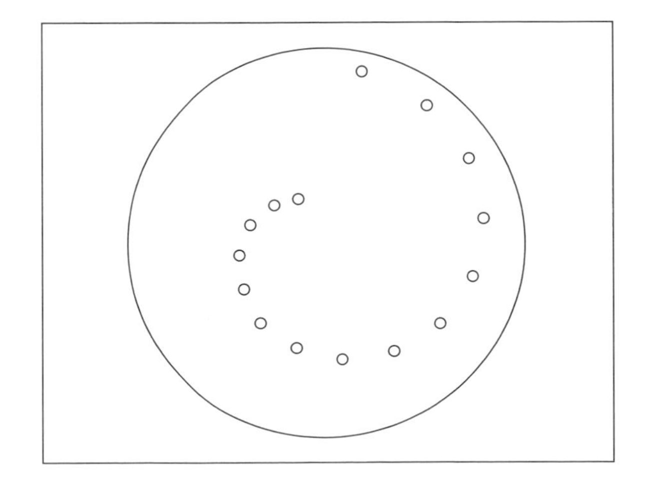

has a set of small holes drilled through a large spinning disk, as shown in the diagram on the previous page. As the disk turns, each hole successively scans a line across a small rectangular area. The scanned line moves steadily down the small rectangle until the pattern repeats. The cumulative effect is that the holes collectively perform what is almost a Cartesian scan over the rectangle. The challenge in our project is thus similar to that faced by Baird and the other television pioneers: to devise a raster scanning system that is easy to implement mechanically.

I have not described any way of synchronizing the two turntables, nor is any such effort necessary. Provided that the two turntables rotate at the same speed, all that synchronizing could do would be to ensure the position of the received image relative to the transmitted image. In fact, the received image is rotating and must be stopped if you want to view it. The angle at which it is stopped can be safely left to the user, who can reorient the image appropriately. What are the effects, however, of small errors in the rotation speed of the receiving disk?

Typically, our helidoodle will transmit a distorted image due to differences in the speeds of rotation of the turntables. Toward the outer edge of the turntable platen, small differences in speed won't make much difference, except that the radial lines in the received image will lean away slightly from the radial (and similarly with other lines with a component resolved along the radial direction). In computerspeak we say that the image has been "morphed" in a simple way, with rectangles turned into parallelograms and circles into leaning ellipses. With larger differences in speed closer to the axis, however, you may see a distortion that morphs radial lines into curves.

And Finally . . . Spirograph Patterns

What other kinds of scans could be devised? The scan figure must cover two dimensions and follow some sort of algorithm easily reproduced by electronics or mechanical means. Could you use an epicycloid curve, as exemplified by the well-known Spirograph drawing machine, for example?

The epicycloid curve of a Spirograph pattern is described by two parametric equations:

$$X = R_h \sin T + R_t \sin(TR_w/R_t)$$

$$Y = R_h \cos T + R_t \cos(TR_w/R_t),$$

where T is a parameter that varies along the scan line, R_t and R_w are the radii of the wheel and track (which would be negative for a wheel rolling around the inside of the track wheel), and R_h is the radius of the hole in the wheel where you put your pen. How efficient could an epicycloid raster scan be? I have tried out a few possible parameters, such as

$R_w = 2$

$R_h = 1$ and

$R_t = -1.1$,

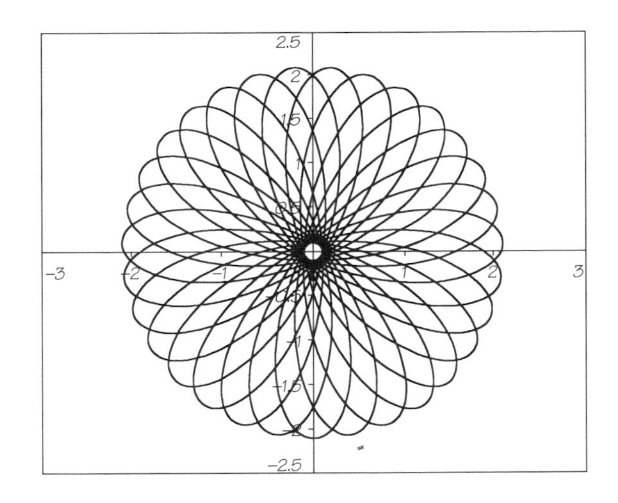

which is the pattern illustrated in the diagram. There seem, though, to be a few potential problems: all the likely-looking scans have many crossovers, and many of the possible rasters concentrate too much around the center or a ring around the center. Perhaps worse, the Spirograph patterns involve a variable speed of motion around the raster figure. The tips of the pattern shown are slow, the parts near the middle much faster. The variable speed of motion could, however, compensate to some extent for the problems of the middle getting too many scan lines. Can you think of a better kind of scan pattern?

REFERENCE

Solymar, Laszlo. *Getting the Message: A History of Communications*. Oxford: Oxford University Press, 1999.

27 Calculator Communicator

> Computers can do lots of things. They can add millions of numbers in the twinkling of an eye. They can outwit chess grandmasters. They can guide weapons to their targets. They can book you onto a plane between a guitar-strumming nun and a nonsmoking physics professor. Some can even play the bongos. That's quite a variety! . . . but at heart they are [all] very similar . . . the innards [are] digital, by digital I mean binary numbers: 1's and 0's.
>
> —Richard Feynman, *Feynman Lectures on Computation*

A vast amount of human communication now takes place on the Internet. And Internet communication employs the language of electronic computers: binary arithmetic. But much of that communication is not really about numbers; it is about human language, most typically written in the characters of an alphabet originally developed in ancient Rome—the alphabet in which this book is written. But the Roman alphabet can easily be expressed in binary numbers by means of a simple code—the ASCII code—and these numbers can then be transmitted. But what advantage can there be in this translation into code? Surely a message that reads

01001000
01100101
01101100
01101100
01101111

is more difficult to manage than "Hello." That may be true for a human processor of information. For a machine, however, it is much easier to handle a 0 or a 1 than to handle the twenty-six lowercase letters, twenty-size capital letters, punctuation marks, and so on—256 characters in all in the ASCII code. A 1 can be represented by a circuit being switched on rather than off, which would be 0. Or a 1 could be a circuit being switched on for 2 nanoseconds, while a 0 would be a 1-nanosecond pulse.

Here we model what happens on the Internet, but at a speed slower, perhaps a million billion (10^{15}) times slower, than what really occurs. Our calculator communicator will run at about one character every 5 seconds, depending upon how quickly you can type the characters.

What You Need

- ❑ 2 printing calculators
- ❑ 4- or 6-way cable
- ❑ 6 6-V relays
- ❑ 6 normally open push-button momentary switches
- ❑ Dongle box in which to fit switches and relays
- ❑ Glue
- ❑ Solder

Alternatively

- ❑ 2 printing calculators
- ❑ 4- or 6-way cable
- ❑ 6 normally open two pole push-button momentary switches
- ❑ Dongle box in which to fit switches
- ❑ Glue
- ❑ Solder

What You Do

The system works by fitting a dongle—an additional little box—to each calculator, which closes the contacts of the keys on both the transmitter and the receiver calculator. With just three keys—1, 0, and Enter—you can send any message you like, in code.

First check that you can get access to the keyboards of the two calculators you are using. Take the casings apart carefully, and gently remove the keyboard

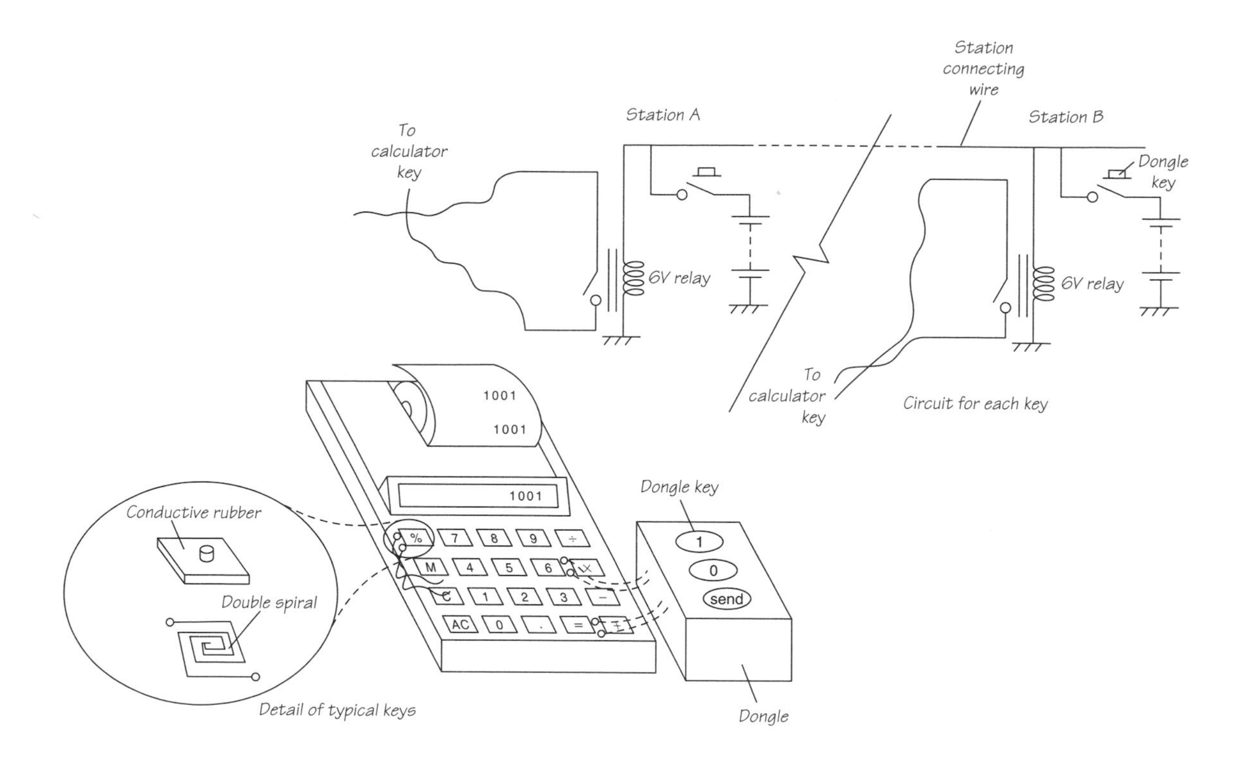

contact membranes and rubber sheets. (These keyboards typically work by having keys molded into a rubber sheet, with conductive rubber pads on the reverse side. When you press on a key, you push the conductive rubber pad onto a plastic membrane, on which are printed interlocking grids of metal conductors underneath each key. The conductive rubber thus functions as a bridge between the two metal conductors, which activate the inputs of the silicon chip's integrated circuit that is the calculator's "brain.")

You should check the wiring and attach wires on either side of the switch points closed by three of the calculator's keys: ideally, the 1 key, the 0 key, and a key like the Enter key that causes a line feed on the calculator printing unit and clears the display for a new number. You may find it convenient to use other keys instead of the Enter key, such as the Divide key, depending upon the layout of the wiring in your calculators. Pull the three sets of two wires out of the casing, perhaps making a groove to accommodate the wires, and then seal the casing again. Ensure that the calculator is working before proceeding. Now wire the six relays as shown, three at each station, with the contacts closing the three key contacts at each calculator, and with the three push-button switches at each end turning on the relays at both ends. Mount all the buttons, relays, and so on inside a small box that you can glue to the side of the calculator. When you press one of the push buttons, it should activate a relay at both ends, which in turn should operate both calculators to produce a 1, a 0, or a line feed/return.

It may be possible to wire the two calculators together, using two-pole switches. This is not the approach I favor, because it may not work at all with some calculators, and it certainly will not work with any calculators over long lengths of cable. The relay solution I have described above works with any pair of calculators over any length of cable.

You may wonder why I suggest using printing calculators, rather than just linking two ordinary calculators. You certainly can use ordinary calculators, and doing so may well reduce the cost of the project. It is useful, however, to have a sequence of printouts in front of you and to write the letter equivalents beside the binary codes on the printouts: you can then read the words coming down the line fairly easily. Working with ordinary calculators involves laboriously writing down the binary numbers or noting only the letter interpretations, with the possibility of errors that cannot be corrected by a second look at the binary. Finally, I think the whole system makes a more enjoyable "toy" when used with printing calculators!

Once all is working, you can try exchanging messages. Remember to use only the three keys on the dongle. You could start with the simplest code set:

a	1	h	1000	o	1111	v	10110
b	10	i	1001	p	10000	w	10111
c	11	j	1010	q	10001	x	11000
d	100	k	1011	r	10010	y	11001
e	101	l	1100	s	10011	z	11010
f	110	m	1101	t	10100		
g	111	n	1110	u	10101		

This binary code is actually a limited-bit version of the ASCII computer code.

THE SCIENCE AND THE MATH

Why Digital?

Why is digital binary transmission such as ASCII a better idea than analog data communications such as the traditional telephone system? The answer lies in the suppression of noise. At low noise levels—for digital systems, especially—small amounts of noise can be suppressed completely using error-correcting codes. In other words, digital systems can yield virtually perfect communication. With analog systems, by contrast, small amounts of noise simply stay there, and they can get worse if multiple retransmissions are necessary.

So how do you suppress the errors in digital signals caused by noise? In principle, this is simple: just transmit more data than you really need at the receiver end. For example, you could transmit the ASCII codes for your data three times and then use the code that got through twice, ignoring a single code set (a sort of "majority rule"). In this way, by transmitting fifteen binary signals, you can be sure to transmit correctly 5 bits of data (a single ASCII character) with perfect reliability if the bits of data are wrong less than one in fifteen times.

In practice, it is possible to devise much more efficient codes than this. For example, by transmitting a 12-bit Hamming code, you can transmit an 8-bit data word correctly with a bit error rate of less than one in twelve, a so-called (12,8) error-correcting code. A simpler example is the (7,4) code below:

0	0 0 0 0 0 0 0
1	0 0 0 0 1 1 1
2	0 0 1 1 0 0 1
3	0 1 1 1 1 1 0
4	0 1 0 1 0 1 0
5	0 1 0 1 1 0 1
6	0 0 1 1 0 1 1
7	0 1 1 0 1 0 0
8	1 0 0 1 0 1 1
9	1 0 0 1 1 0 0
10	1 0 1 0 0 1 0
11	1 0 1 0 1 0 1
12	1 1 0 0 0 0 1
13	1 1 0 0 1 1 0
14	1 1 1 1 0 0 0
15	1 1 1 1 1 1 1

Try changing any single bit in any one of the codes, and you will see that it is still recognizably that code and not one of the others. Here are the possible errors in code 4, for example:

4	0 1 0 1 0 1 0	model
4	1 1 0 1 0 1 0	error in position 1
4	0 0 0 1 0 1 0	error in position 2
4	0 1 1 1 0 1 0	error in position 3
4	0 1 0 0 0 1 0	error in position 4
4	0 1 0 1 1 1 0	error in position 5
4	0 1 0 1 0 0 0	error in position 6
4	0 1 0 1 0 1 1	error in position 7

None of these single-bit error transmissions of the code for number 4 can be confused with the codes for 0, 1, 2, or 3, or any of 5 to 15.

Why Do You Need Relays to Connect Calculators?

Connecting calculators is not straightforward because they interfere with each other's operations. The keyboard switches do not simply provide an input to the calculator; they also carry the voltages of an output from the calculator circuits—and these outputs are not synchronized, which generally causes malfunctions on both of the devices that are connected.

Instead of individually connecting all twenty-five or thirty keys to the processor, they are typically connected by means of an *x-y* grid, so that only ten or eleven lines are needed from the keyboard into the processor. (It is expensive to link lots of wires to a processor chip, so this technology makes calculators cheaper.) Calculators have a "poling" or scanning system that takes inputs from this keyboard grid into the central processing unit. They output voltage pulses on the *x* lines in succession, and check whether the voltage of any of these pulses appears on any of the *y* lines. This scanning takes place at high speed, to ensure that no pressing of a key, even the shortest, is missed.

You can deal with these problems, at least for short lengths of cables, by cross-connecting the calculators with double-pole push-button switches. This approach avoids the possibility of one calculator's outputs appearing on the internal wiring of the other. The poling technology prevents cross-wiring using two-pole switches on long lines, however, which is why I suggest using relays instead.

Different calculators employ different scanning methods. However, you can obtain the general idea from reviewing the data on a typical keyboard scanner integrated-circuit chip such as the MM74C922 from Fairchild. This device scans the keyboard's four x lines with a clock speed from 30 to 6,000 Hz, so to scan four lines takes between 130 and 0.7 ms. The outputs are registered on the 5 y lines, which lead to the decoding circuit, where they are converted to a binary output between 0 and 19.

At the highest speeds, pulses on the different lines will be distorted or destroyed by the large capacitance value of the cable, which will cause malfunctions. If we assume a line length L of 10 m, and a coaxial cable with a 1-mm core and 0.5-mm insulation, then we can calculate the capacitance C as follows:

$$C = \varepsilon\varepsilon_0 Ld/t,$$

where ε is the dielectric constant for the material, which might, for example, be 3.5 for PVC. Hence,

$$C \sim 3.5 \cdot 8.85 \times 10^{-12} \cdot 10 \cdot 0.001/0.001 = 300 \text{ pF.}$$

With a resistance of the driving lines of, say, 1 MΩ, the time constant τ of the cable will be

$$\tau = 1 \times 10^6 \cdot 300 \times 10^{-12} = 0.3 \text{ ms.}$$

This limitation is enough to cause problems, given the sharp edges needed for the narrow pulses to function correctly.

And Finally . . . Binary Morse Code

You may wonder why ASCII-type binary coding is used in today's computers rather than Morse-based coding. Morse should have had a fundamental advantage, because it saves time by sending short codes for the more commonly used characters. Some of the answer is historical accident, which I won't address. Some of the answer, however, lies in ASCII's convenience of use inside a computer, where data moves around on sets of parallel wires, quite typically eight wires, sometimes sixteen or thirty-two wires. If you consider the eight-wire example, you can understand how an ASCII byte fits nicely across those wires. Using Morse would not save any time, because even an E or a T, represented by a single 0 and single 1, perhaps, would still be transmitted on the eight wires in parallel, with the other seven wires doing nothing.

Once the code has escaped from a computer and is transmitted across the Internet, there would be advantage in the Morse code. But actually we can use a highly complex and capable machine—the computer—to re-encode the ASCII in a compressed-data form. For example, at the character level, we can suppress the transmission of a string of seventy-six identical characters by transmitting a special multiplier code, followed by the character to be multiplied. And we can re-encode and compress at the binary level by looking for long strings of 1s and 0s.

What about devising a kind of Morse code for the binary calculator communicator? You couldn't use it directly, coding a dot as a 0 and a dash as a 1. This is because you cannot transmit double 0s. Thus, although a single 0 for a Morse E would be okay, you couldn't transmit 00 to represent Morse I, or 001 to symbolize Morse U. How else could you use the Morse principle?

And what about a code that used a third number or letter on the keyboard? This would be a ternary digital communications system, which might offer some advantages. It has been shown, for example, that computer circuits could be built using a ternary number system, and these could equal or perhaps even improve the performance of binary circuits (see Epstein, *Multiple-Valued Logic Design*). You could try this on the calculator communicator by using the keys 0, 1, and 2. What advantages and disadvantages can you discover?

REFERENCES

Epstein, George. *Multiple-Valued Logic Design: An Introduction.* Bristol, U.K.: IOP Publishing, 1993.

Singh, Simon. *The Code Book: The Science of Secrecy from Ancient Egypt to Quantum Cryptography.* London: Fourth Estate, 1999.

Instruments

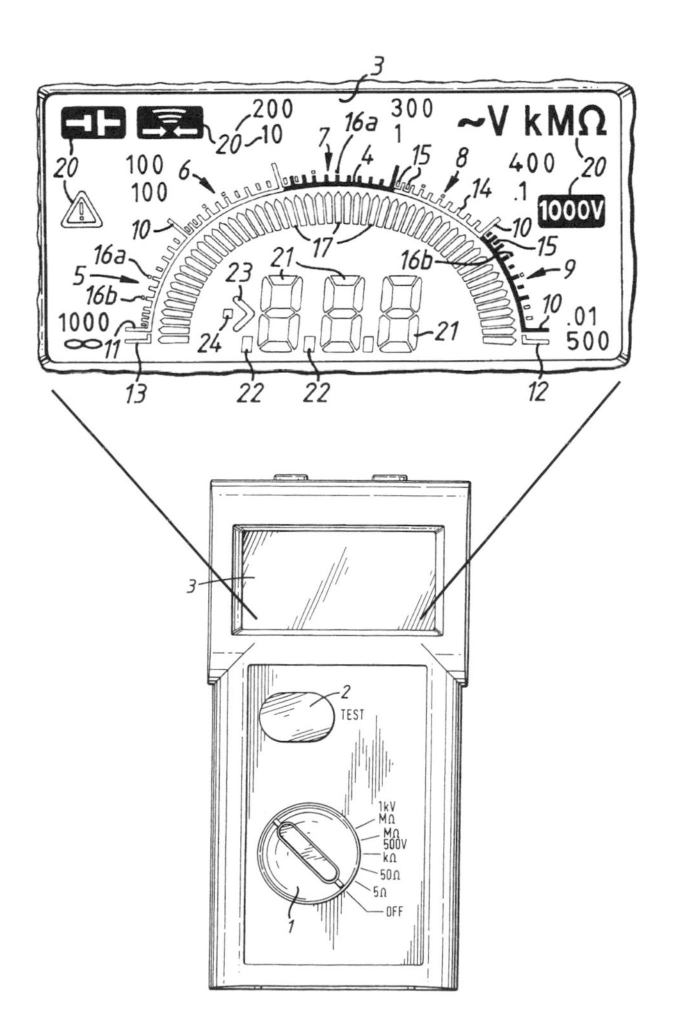

Reports that something hasn't happened are interesting to me because as we know there are known knowns—there are things that we know that we know. There are known unknowns—that is to say, there are things that we now know we don't know. But there are also unknown unknowns—there are things we do not know we don't know. And each year we discover a few more of those unknown unknowns.

—Donald H. Rumsfeld, U.S. secretary of defense, press briefing, February 2002

Science is about the discovery of unknowns, and then, most importantly, measuring them. Measurement—making knowledge quantitative—usually relies on the objective power of instruments. Without some instruments, we wouldn't be able to sense many things. But even when we can use our senses, an instrument is usually needed to give us quantitative knowledge.

Our first instrument addresses a quantitative question: not whether you stretch when you get out of bed in the morning, but rather how much you stretch. An impossible question to answer—unless you have a posture meter to tell you. After that, we attempt to analyze material that we can neither touch nor see: even a qualitative answer is impossible where the analysis of gases is involved, unless you have an instrument. For our first gas analyzer, we modify a household smoke detector, while for the second, we explore the curious ability—not encountered in ordinary gases or at ordinary frequencies—that some gases have to absorb sound waves. Finally, we consider what must be two of the world's simplest instruments: a tiny glass of water and a small piece of black plastic. What can these possibly tell us about the world? Read on. . . .

28 Posture Meter

> Are you sitting comfortably? Then I'll begin.
>
> —Julia Lang, on BBC's *Listen with Mother,* 1950–82

What constitutes a comfortable position depends upon ambient conditions: if it is hot, then a comfortable position may be stretched out; in a cold environment, curling up may seem more comfortable. A constant posture is also undesirable for a human being: we need to change our posture every few minutes, even when sleeping. Speeded-up videos of people asleep show them wriggling, rolling, stretching, and curling up all night. Maybe we could monitor people's posture changes during the night more simply with a posture meter. But what is a posture meter, and how does it work?

What You Need

- ❑ Multimeter with capacitor measurement function
- ❑ Insulating plate
- ❑ Wire with alligator clips
- ❑ Ground connection

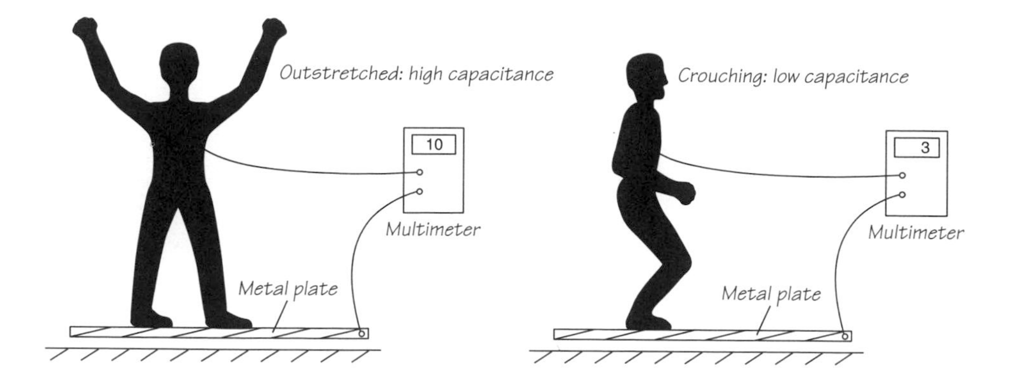

What You Do

Here we measure a single parameter—oddly, an electrical parameter: the electrical capacitance of the human body with respect to the earth—and show how it is related to posture. There are two ways to demonstrate the posture meter: the use of induced voltage from the standard electrical power, and the use of a multimeter with a readout designed to measure capacitance.

With a Capacitance Meter

First, connect wires with alligator clips to the multimeter's two capacitor-measuring leads. Connect one of the alligator clips to a ground—a large metal item, ideally connected to the water system or to the ground of your electricity supply. Take your shoes and socks off and stand on the insulating plate with the other alligator clip between your toes. Now squat down, raising first one arm, then the other.

If the multimeter readout is sufficiently sensitive, the reading will vary with your posture. Stand up and stretch your arms in the air, and you'll get a large reading. Squat down with your arms crossed, and you'll get a small reading. Ask different people to try out the posture meter, perhaps starting everyone from a standard standing or stretched-out posture.

A multimeter typically measures capacitance by applying a small audio-frequency AC voltage to the unknown capacitance and then measuring the average current. Alternative schemes involve applying a brief voltage pulse and measuring the decay time at two points: the more the decay, the less the capacitance. The multimeter I used followed the latter scheme: the test used 0.1 volt pulses at a frequency of 10 Hz.

The 50/60-Hz Pickup Method

This method uses a multimeter or voltmeter without any special capability for capacitance measurements. Connect one lead to the ground and one to the plate on which you stand. Switch the meter to a fairly sensitive AC scale, probably 200 millivolts or less. Now stand, crouch, and stretch your arms out, and your voltage will change with your posture.

How It Works

The posture meter works because the human body is an antenna, whose reception efficiency depends upon its capacitance to ground—exactly the same quantity we measure directly with the capacitance meter. The more capacitance, the larger the 50/60-Hz pickup signal, and vice versa. The antenna in this case is not receiving a radio signal as such, but rather the accidental emission of domestic power lines, which radiate minute amounts of power at 60 or 50 Hz and the harmonics of those frequencies. Assuming that the amount of radiated power does not change, we can thus measure capacitance by measuring the amount of induced voltage.

The capacitance C of a flat-plate capacitor is the amount of electric charge Q that can be stored at a certain voltage V:

$$Q = CV.$$

It is proportional to the area of the capacitor divided by the distance, t, between the plates,

$$C = eA/t,$$

where e is a constant. Proportionality to the plate area A is not a surprise; imagine putting more capacitors together in parallel and you can understand that the charges on the plates of each of them would add up. But why the inverse proportion to the distance between the plates? Separating two charged plates will increase the voltage because it decreases the capacitance, according to

$$V = Q/C = Qt/eA,$$

in other words, the voltage is proportional to the distance the plates are apart. Clearly, plates at higher voltage, with the same charge, mean more energy. Where is this energy coming from? It is coming from the energy needed to pull the charged plates apart.

It is a little difficult to apply the idea of the flat-plate capacitor to something the shape of the human body; but it does seem to make sense that, when we stretch out, we will have a much higher capacitance than when we are curled up. Basically, we are presenting a greater area to be influenced by voltage. Below we look more precisely at the math behind these observations.

THE SCIENCE AND THE MATH

The capacitance of a flat-plate capacitor C can be expressed as

$$C = \varepsilon_0 A/t,$$

where A is its area, t is the plate separation, and ε_0 is the permittivity ("dielectric constant") of free space, which is about 8.85×10^{-12}F. Its exact value is $1/(4\pi c^2 \times 10^{-7})$ F, where c is the speed of light.

Two concentric spheres, one of radius R on the outside and R' on the inside, have an area approximately equal to that of the larger sphere, and a separation roughly that of the difference in radius, so

$$C = \varepsilon_0 A/t = \varepsilon_0 4\pi R'^2/(R' - R).$$

This relationship will tend to $4\pi\varepsilon_0 R'$ as R approaches infinity, so you will find it plausible that the capacitance of an isolated sphere C is in fact

$$C = 4\pi\varepsilon_0 R.$$

When you stretch your arms above your head, you are simulating a sizable increase in "volume," as far as capacity goes. We can understand this if we use the equation for the capacitance of a prolate ellipsoid* (American or rugby football shape), in which the major axis is $2a$ long and the minor axes are $2b$ long. Now

$$C = 4\pi\varepsilon_0 (ab^2)^{1/3}\, x^{-1/3} \sqrt{(x^2 - 1)}/\log_e(x + \sqrt{(x^2 - 1)}),$$

*You can remember what a *prolate* ellipsoid is by imagining a *prol*ix person—someone who talks long-windedly—with an enormously long, elliptical speech bubble coming out of his or her mouth. For an *oblate* ellipsoid (M&M/Smarties candy shape), imagine a comically *ob*ese person: someone whose waist diameter exceeds his or her height.

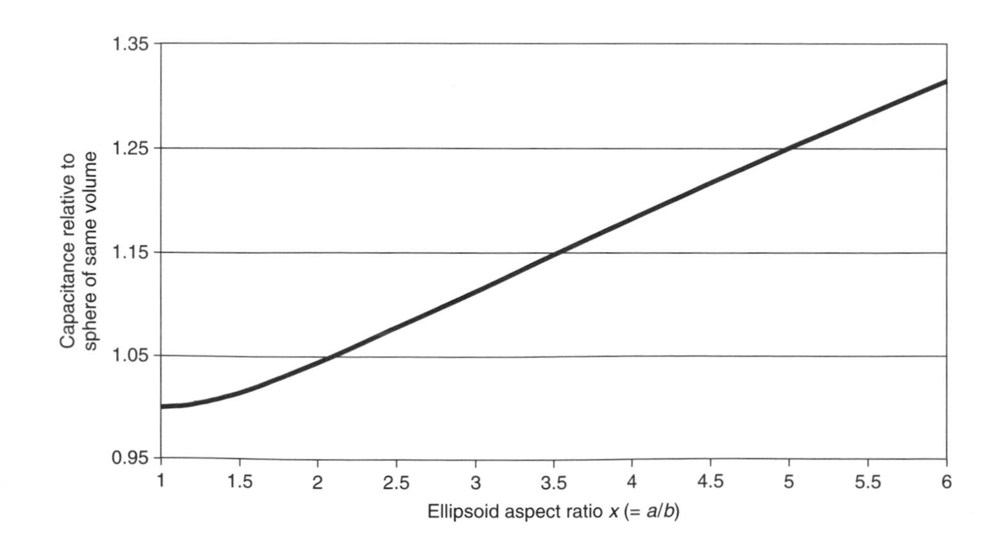

where $x = a/b$ gives the capacitance of a prolate ellipsoid. This can be alternatively expressed as

$$C_{rel} = x^{-1/3}\sqrt{x^2 - 1}/\log_e (x + \sqrt{x^2 - 1}),$$

where C_{rel} denotes capacitance relative to a sphere of the same volume.

The capacitance of a sphere increases linearly with the radius of that sphere. This function for an ellipsoid increases more slowly than a linear increase of capacitance with the height of the person stretching.

An ellipsoid reflects the shape of a person stretching with arms together more than with arms spread wide. Connect the meter again, and you can easily discover that you register a smaller capacitance when you put your hands together above your head, better simulating a prolate ellipsoid. The graph plots C_{rel} versus x. At large x, the formula can be stated as

$$C_{rel} = x^{2/3} / \log_e(2x),$$

which clearly shows the relation to the spherical formula: as x increases, the capacitance increases. In other words, stand up and your capacitance will rise.

And Finally . . . a Weight Loss Meter

Rather than measuring the capacitance value at low frequency, you could try measuring capacitance value at different frequencies versus body posture.

Could the posture meter be of any practical value? Could it be used by physicians, for example, to judge whether patients need to lose weight? There are devices that, by comparing weight to an electrical conductivity measurement, aim to do just that, although the results are only approximate. Could our capacitance method be better?

29 No Smoke without Gas

What is claimed is:

1. Element 95.
2. The isotope of element 95 having a mass number of 241.

—Glenn T. Seaborg, U.S. patent no. 3156523, 1964
(one of two elements to have been patented)

The smoke detector is a great invention because it does good. Thousands of people owe their lives to smoke detectors that enabled them to escape fires that started in their houses while they were asleep. Many more people have smoke detectors to thank for providing prompt warnings of accidental or incipient fires, averting damage to both buildings and properties.

The smoke detector is also a great invention because it is ingenious. The large market for detectors and the many competing suppliers, allied to the magic of mass production, mean that we get a large chunk of technology and skillful design for a small amount of money. As someone with an economical turn of mind, I have long been intrigued by the possibility of using one of these bargain devices for something different.

Different smoke detectors reflect varying principles. The most common detectors are based on absorption of a light beam by smoke or on the reduction of current in an ionization chamber. We use the latter type of device in this project. However, we don't use the smoke detector to detect smoke. Instead we redeploy it as a gas sensor that can indicate at least roughly the concentration of a gas in the air.

What You Need

- ❑ Battery-powered smoke detector of the ionization type (e.g., with the Motorola chip MC14467)
- ❑ Multimeter
- ❑ Lunch box with sealable lid
- ❑ Large (1- or 2-liter; 2- or 4-pint) plastic vessels (e.g., beakers or measuring jugs)
- ❑ Flexible plastic tubing, 10 mm (3/8 inches) in diameter
- ❑ Sample gases (e.g., helium balloon, CO_2 soda siphon, HFC134a freezer spray)

What You Do

A smoke sensor contains a small radioactive source that slightly ionizes the air inside. A small voltage is applied, which causes the charged air molecules—ions—to pass a minute current through the air. If smoke enters the sensor cham-

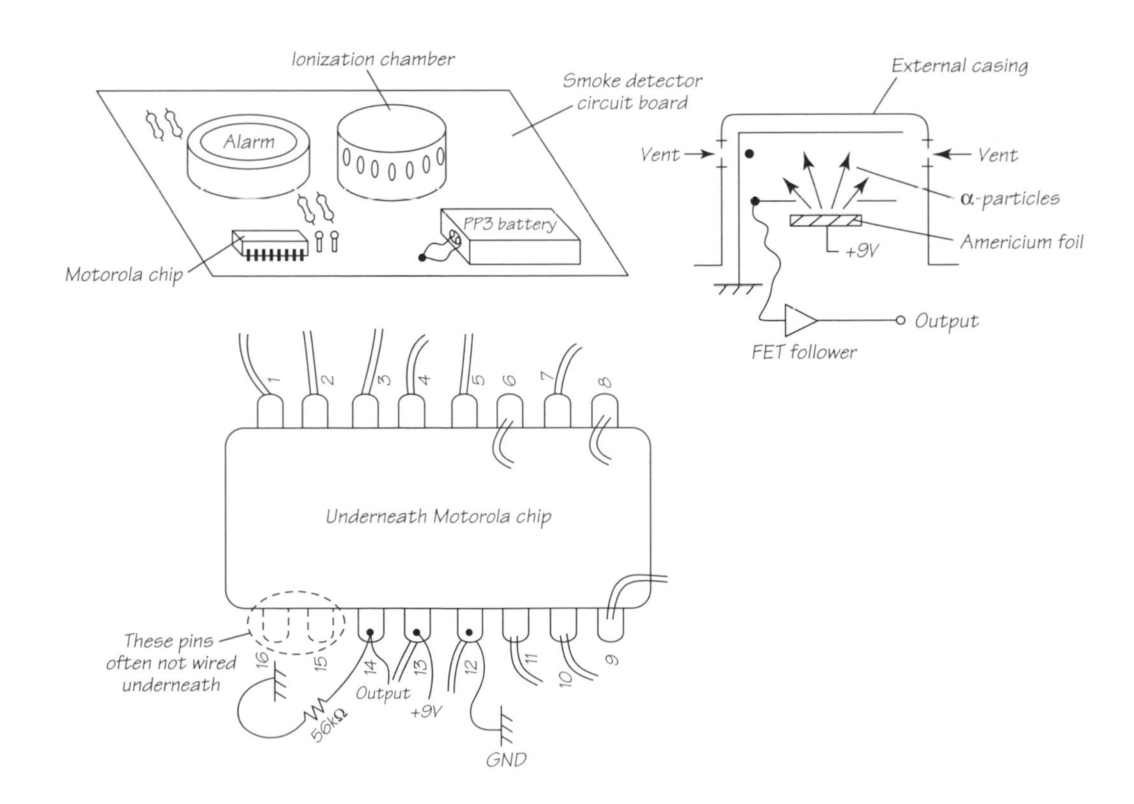

ber, the smoke particles attract the charged air molecules and become charged smoke particles. However, the smoke particles are much larger, between 10 and 1000 nm in diameter—at least twenty times larger than the air molecules in diameter and 10,000 times larger in mass. These massive particles don't carry current effectively in a small electric field. So the measured current falls, and this drop in current triggers the alarm bell.

Different gas molecules inside the sensor will also affect the current, because their masses differ from those of air molecules. Although not as dramatic a change as with smoke, this is an effect we can measure.

First, carefully dismantle the cover of the smoke detector. Rewire the circuitry according to the diagram and the following discussion. (You can download technical data describing the details of the Motorola integrated circuit MC14467 from the Motorola Web site and other places on the Internet.) Basically, we are rewiring the device so that it will not sound the alarm and so that it does not go into a "sleep mode," which it routinely uses to save battery power. In effect, the device becomes a low-impedance follower circuit for the high-impedance voltage output of the ionization chamber assembly.
Specifically:

- We connect a 56k resistor to ground from pin 14, to suppress "power hum" and other interference effects caused by a high-impedance circuit.
- Pin 12 is wired to ground to keep the power on all the time.
- Pin 13 is wired to +9 V to keep the alarm off.
- Additionally, you can cut the alarm wires (there are usually three) and remove the alarm unit, thus making the sensor more compact.

WARNING

Do not attempt to take apart the smoke detector's ionization chamber. If you do, the minute amount of contamination from your fingers will destroy the chamber's function by spoiling its high insulation resistance. The chamber also contains a tiny radioactive source, about 1 microcurie of americium-241. This substance is not dangerous outside your body—the alpha particles will not penetrate your skin to the living cells beneath. If you were particularly unlucky in handling the inside of the chamber, however, you just

Alpha particles ionize the air inside the chamber, providing a number of ion pairs like N_2^+, N_2^-, and the like. With voltage applied, the ions drift to the chamber plates, where they contribute a minute current. This minute current flows through two chambers—the detection chamber and a ballast chamber—wired so that they form an electric potential divider. An ultra-high-input impedance field-effect transistor (FET) amplifier then follows the voltage of the potential divider. It is this voltage that we bring out to the multimeter, using the amplifier in the Motorola integrated circuit.

Now connect the modified sensor to a normal high-impedance (digital) voltmeter or multimeter. The output will be something like 3 to 6 volts, wobbling up and down 0.1 volts or less, perhaps jumping around more when you move the sensor around.

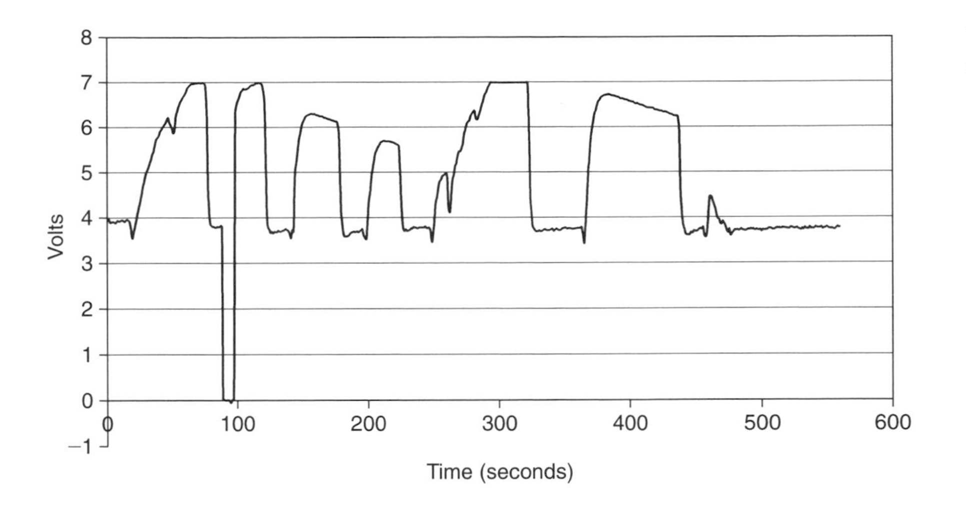

WARNING

might swallow some of the radioactive source, in which case it would be dangerous (although less dangerous by virtue of the minuscule amount involved).*

The easiest test gas to use with the sensor is probably a hydrofluorocarbon 134a "freezer spray." HFC 134a is a non-ozone-depleting, low-global-warming, liquefied gas used as the propellant in many small aerosol spray cans and pressurized medical products. Freezer spray is commonly just HFC 134a, the pure propellant without any liquid product to propel. It evaporates quickly on leaving the nozzle and can cool things down to near its boiling point of –26 °C. The formula for 134a is 1,1,1,2-tetrafluorethane:

CF_3-CH_2F

and thus has a molecular weight of 102, appreciably heavier than air with an average molecular weight around 29. You should find that the sensor will increase in readout with 134a, by about 3.2 volts in the case of the sensor I tested.

An alternative test gas you may have around your house is butane, which contains two isomers:

CH_3-CH_2-CH_2-CH_3 (normal butane)

and

CH_3-CH-CH_3 (isobutane)
\
CH_3

Butane yields a smaller response than HFC 134a, as you might expect from its lower molecular weight (58), around +2.9 volts in the case of the sensor I tested.

*This is what the U.S. Environmental Protection Agency says about 1-microcurie Am-241 ionization smoke detectors: "As long as the radiation source stays in the detector, exposures would be negligible (less than about 1/100 of a millirem per year), since alpha particles cannot travel very far or penetrate even a single sheet of paper, and the gamma rays emitted by americium are relatively weak. If the source were removed, it would be very easy for a small child to swallow, but even then exposures would be very low because the source would pass through the body fairly rapidly (by contrast, the same amount of americium in a loose powdered form would give a significant dose if swallowed or inhaled). Still, it's not a good idea to separate the source from the detector apparatus."

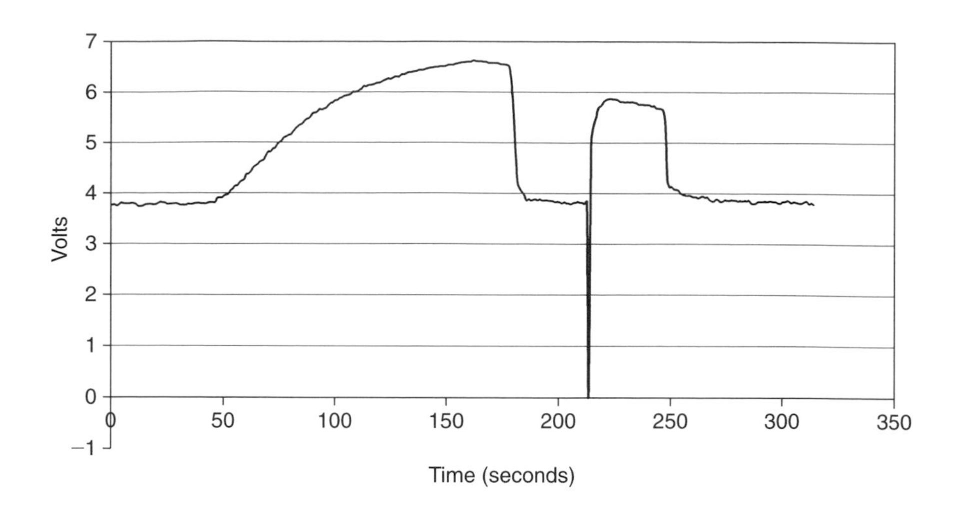

WARNING

Butane is flammable and can form explosive mixtures with air. Keep away from any sources of ignition, and work outside with a plastic beaker. The easiest way to dispense butane is to block up the air holes of a small, handheld plumber's burner with tape and then to allow gas from the burner to flow down the side of the beaker.

These are both heavy gases, so you can treat them a little like you might water, "pouring" them into the bottom of a beaker and displacing the air upward. The two plots show the readouts from a sensor I made in a 1-l plastic beaker. In each case, I slowly squirted in gas at the bottom of the beaker (the time on the x-axis is in seconds), pulled the sensor out of the beaker, and then replaced it, showing the sensor's fast-step response and also showing how the gas is diluted with air when the sensor is put back into the beaker. The HFC 134a plot shows that I did this twice, to see whether the glitch to zero volts recurred (it did not; it was probably due to a momentary short circuit). The second plot shows the butane response.

If you have a helium balloon handy, try that with the sensor. You will need to turn your 1-l beaker upside down, however, and gently fill the beaker from the top down from underneath, because helium is lighter than air. (Imagine that the ceiling is the floor and that you are pouring water into the beaker.) You should find that the sensor will indicate a drop; in the sensor I tried, it registered a change of –2V.

JAMES LOVELOCK, THE ELECTRON CAPTURE DETECTOR, AND THE GAIA THEORY

The gas chromatograph (GC) has been the workhorse of much chemical analysis for the past sixty years. A GC separates a gas or vaporized liquid into its components and then routes them one by one past the same detector. The peaks in a plot of the detector's output versus time tell you what the gas mixture contained.

Early GCs were not particularly sensitive. They relied on physically sensing the gases, using effects like thermal conductivity, which vary only by a factor of 1 to 10 between gases. As a result of biomedical studies in the 1960s, James Lovelock devised a simple but much more sensitive detector, the electron capture detector (ECD). This is a selective detector that records a massive response to halogen elements such as chlorine or fluorine in vapors but responds little to other gases.

The ECD is remarkably similar in concept to our smoke-alarm Am-241 absorption detector, since it also involves the use of a radioactive source and ionization. The Ni-63 source produces beta particles; it is much more powerful than our Am-241 source, and the gas chromatograph to which it is normally connected concentrates the gas sample to be measured. Many of the differences between the ECD and our crude device stem from the way in which the ECD responds selectively to electron-absorbing (i.e., halogen-containing) artificial molecules. This selectivity arises because the ionization caused by the Ni-63 produces a flood of low-energy electrons that can be captured easily by the electrophilic halogen molecules. The result of this selective ionization makes a huge difference: whereas our device can barely measure a change in gas concentration of 2 percent, the ECD can pick up concentrations a trillion times smaller.

The ECD enabled the measurement of tiny traces of artificial molecules, such as insecticides, down to parts-per-trillion levels or less (1 part in 10^{12} to 1 part in 10^{16}). This rather technical invention also gave rise to a new power in politics: the modern environmental movement. For the first time, it became possible to monitor the progress of artificial chemicals not just from factory to insect or weed, but also, at increasing dilutions, to the soil on farms, along watercourses, into animals that eat insects, and even into the carnivorous animals (such as ourselves) that ate the insectivores. Rachel Carson wrote her book *Silent Spring,* with its solemn warnings about environmental catastrophe, and a whole new global political movement was born.

James Lovelock later devised the Gaia theory, which posits that the whole of the earth's surface, including land, oceans, and atmosphere, and all the plants and animals, is a kind of collective superorganism. This superorganism, which he called Gaia, reacts against changes in order to maintain itself, demonstrating on a vast scale what a biologist would call homeostasis. His theoretical model, "Daisy World," showed how life could regulate temperature. Daisy World's simplified Earth is a precisely spherical planet populated only by daisies whose color varies from white to black.

Although a hero of the environmental movement because of his ECD device, and an ardent environmentalist himself, Lovelock also stood up for what is scientifically respectable. His Gaia theory actually rejects many of environmentalists' less watertight arguments about the dangers of pollution and global warming, although that doesn't let us off the hook: humankind is certainly still creating big environmental problems that will require solutions.

Gaia theory is still controversial, but that controversy has generated a lot of good science and greatly boosted our knowledge of ecology and our environment. Gaia theory highlighted the interactivity and interdependence of all life on the earth. It also highlighted the importance of processes that were previously considered laboratory curiosities—for example, the emission from bacteria of minute amounts of exotic and poisonous chemicals such as hydrides and alkyls.

Americium is an artificial metal produced by the modern alchemical transmutations of the elements that occur in nuclear reactors. All of americium's isotopes, from Am-237 through Am-246, are radioactive. Am-241 is formed spontaneously by the beta decay of plutonium-241, which is itself formed by two successive neutron absorptions. Trace quantities of americium are widely used in smoke detectors and as neutron sources in neutron moisture gauges.

Am-241 is an alpha-particle emitter:

Am-241 $\rightarrow$ Np-227 + He-4.

Ionization-sensor smoke alarms contain a small amount of Am-241 embedded in a gold foil within an ionization chamber. This source is made by rolling gold and americium oxide ingots together to form a foil only a few micrometers thick. This thin gold-americium foil is then sandwiched between a 250-micron silver back and a 2-micron-thick palladium cap layer. The palladium is a corrosion-proof layer (palladium is a platinum group metal) thick enough to retain the americium completely, but thin enough to allow the alpha particles to pass. Am-241 has a half-life of 470 years, which means that even over the longest conceivable lifetime of a smoke detector—perhaps fifty years—it will not have decayed by much. A shorter-lived isotope would lead to a shift in the detector's output voltage because of its declining activity and so would not be as reliable.

The ionization chamber comprises two or three metal plates a small distance apart. One of the plates carries a positive charge, the other a negative charge. Between the plates, air molecules—oxygen and nitrogen—are ionized when electrons are ejected from the molecules by the larger, heavier alpha particles from the radioactive material. The presence of the ions allows a small electric current to flow when there is a voltage on the plates. In the most common design, the dual-chamber ionization smoke sensor, there are two ionization regions between the top, grounded plate and the bottom, positive, alpha-emitting plate.

In clean air, the ionization and the electric fields are such that an equal current flows from the top plate to the middle and from the middle to the top plate: the middle plate is thus kept at a voltage intermediate between ground and +9 V. When smoke enters the ionization chamber, the current is disrupted as the smoke particles attach to the charged ions and restore them to a neutral electrical state in the larger, open part of the chamber. This reduces the flow of electricity between the two top plates in the ionization chamber, increasing its "resistance," while the other part of the chamber is much less affected. The result is that the voltage on the middle plate increases toward +9 V. When the voltages reach a certain threshold, the alarm—the function we have disabled in our project—is triggered.

Smoke ionization detectors are most sensitive to fast-flaming fires, which are characterized by combustion particles in the 0.01- to 0.3-micron range. (Photoelectric smoke detectors are more sensitive to slow, smoldering fires, which are characterized by a preponderance of larger particles in the 0.3- to 10.0-micron range.) These submicron particles do most of the ionization current reduction. In our case, the sensor works in exactly the same way, but we vary the ionization current because of the gas we are adding.

We can estimate roughly how much current the ionization chamber conducts when it is operating correctly in air, if we know the power of the source. The source has a power of 1 microcurie, so we have 3.7×10^4 disintegrations per second, and these produce alpha particles with an energy of 5.5 MeV.

If we assume that the ion formation energy is of the same order as the dissociation energy of a diatomic molecule such as O_2 (494 kJ/mole) or N_2 (942 kJ/mole), then we can see that a single ion pair will consume an energy of

$$700{,}000 / 6.023 \times 10^{-23} = 1.2 \times 10^{-18} = 7 \text{ eV}.$$

If each alpha particle produces ion pairs with 50 percent of its energy, then

$0.5(5.5 \times 10^{6})(3.7 \times 10^{4})/7 = 1.45 \times 10^{10}$ ion pairs per second.

Most of these ion pairs are probably collected by the plates in the absence of smoke, so the current will be approximately

$$2 \times 1.45 \times 10^{10} \times (1.6 \times 10^{-19}) = 5 \text{ nA}.$$

Although this current may sound small, it is the upper limit of how much current could possibly be produced: the ion formation and collection process are actually reduced by geometrical factors. Alpha particles emitted in the wrong direction, or which are not completely stopped within the air in the chamber, will not contribute to the current. This tiny current, or even smaller currents, can, however, easily be measured by an ultra-high-impedance measuring circuit: the current is arranged to flow down an extremely large resistor, and the voltage drop across this resistor is measured. With a 10^{-9}-ohm resistor in the way, a 1-nA current will give a clearly detectable 1-volt signal.

We are not quite finished. If you connect an ordinary voltmeter to try to measure that 1-volt signal across the 10^{9} resistor, it will probably indicate nearly zero. This is because most voltmeters take an appreciable current—typically a fraction of a microamp (about 10 microamps for a moving coil meter)—from the circuit they are measuring. An FET integrated-circuit buffer amplifier is needed, which will present that voltage to the measuring circuit without drawing any current from the measured circuit. An FET integrated-circuit operational amplifier can have an effective input impedance as high as 10^{14} ohms. Thus, at 10 volts, only 10^{-13} amps would flow into the amplifier. Current flows of the order of picoamps (10^{-12} A) can therefore be measured accurately by a simple electronic amplifier of the right design, using an FET.

And Finally . . . Stabilizing the Ionization Chamber

Can you improve the stability of the ionization chamber somehow? A simple resistor/capacitor filter circuit would improve matters, for example. If you wired two chambers together to generate a signal that was an average of the two, how much would the signal improve?

REFERENCES

Braithwaite, A., and F. J. Smith. *Chromatographic Methods.* 4th ed. London: Chapman and Hall, 1985.

Lovelock, James. *Gaia: A New Look at Life on Earth.* 1979. Reprint with corrections and a new preface. Oxford: Oxford University Press, 2000.

30 Spongy Gas

> It was a foggy day in London, and the fog was heavy and dark. . . . Gaslights flared in the shops with a haggard and unblest air, as knowing themselves to be night-creatures that had no business abroad under the sun; while the sun itself when it was for a few moments dimly indicated through circling eddies of fog, showed as if it had gone out and were collapsing flat and cold. . . . the loftiest buildings made an occasional struggle to get their heads above the foggy sea, and especially that the great dome of Saint Paul's seemed to die hard; but this was not perceivable in the streets . . . where the whole metropolis was a heap of vapour charged with muffled sound of wheels.
>
> —Charles Dickens, *Our Mutual Friend*

If you shout, even from the rooftops, the whole world won't hear you. In the open air—in a field, for example—the human voice carries only a couple of hundred yards at best, and in a London fog much less. You can extend this range a little by shouting over a hard surface—one that reflects sound waves efficiently—which is one reason why military drill instructors sound loud. Whales can't shout to the whole world either, but they can get much closer to it. Whales "sing" to each other underwater, and the sound can carry several hundred kilometers. So why is it that in air the distances over which sound will travel are much more limited?

Shouting range is not normally limited by an absorption process: not much sound is converted into heat or molecular energy. Rather, sound is reduced by dilution or refraction. Dilution of sound follows something like the well-known inverse square law followed by all radiations. Occasionally, loud explosions are heard at long distances—10 miles or more—so there is nothing inherently limited about the range of sound.

Deflection of sound waves by refraction—bending—is the other principle limiting sound range. Refraction is a familiar phenomenon: on hot, sunny days the pace of life seems slower because sounds are muffled. Traffic noise, construction sounds, all these fade, seeming to recede to an immense distance. This is because the air near the ground is hotter than the air higher up, and sound waves from ground-based sources are refracted in an upward arc rather than traveling sideways. Take a ride on a hot-air balloon, and you will find where some of that sound has gone: upward. Aficionados of ballooning will tell you how one can hear ordinary quiet conversations on the ground from even hundreds of feet aloft.

All this might suggest that sound should go on forever, that in ideal circumstances a "shot heard 'round the world" should be possible in the right circumstances—if it's loud enough to deal with the inverse square law. But such is not the case. At least in polyatomic gases, or in air containing appreciable quantities of them, at long distances absorption does eventually become important. Sound is not simply diluted and refracted—even in the most ideal situation—but is absorbed as well. Furthermore, as we shall discover, at high frequency sound can be absorbed more rapidly—sometimes in as little as 3 feet or less.

What You Need

- ❑ 40-kHz ultrasonic transducers (often sold in pairs, sometimes marked 40T for the transmitter and 40R for the receiver)
- ❑ Transistors, capacitors, potentiometer, and resistors for oscillator as per circuit
- ❑ Spongy tubing (pipe insulation material), inner diameter 25 mm (1 inch), outer diameter 60 mm (2.5 inches), about 1 m (3 feet) long
- ❑ Oscilloscope or AC millivoltmeter (a function on more sophisticated multimeters)
- ❑ Carbon dioxide (available in soda siphons and Sodastream machines)
- ❑ Hydrofluorocarbon 134a (used in "freezer sprays" and in many medical aerosols)
- ❑ Butane (barbecue gas; see safety warning)

For audio-modulated version

- ❑ Transistors and diodes for modulator and detector circuit as per circuit
- ❑ Audio amplifier
- ❑ Loudspeaker

What You Do

The objective of the setup in the diagram is to make an ultrasonic absorption system. It comprises an emitter and receiver for ultrasonic waves with a volume into which different gases can be injected. The emitter and receiver are similar resonant piezoelectric transducers. The emitter needs an electronic oscillator feeding it, while the receiver may need an amplifier.

Having assembled a simple multivibrator oscillator circuit, you need to tune it so that it will operate at a resonant frequency of the ultrasonic transducer. To do this, connect the receiver transducer to the AC millivoltmeter and turn it to the highest sensitivity, or until you see a signal. (If power hum—AC voltages being measured even when nothing is connected—proves a problem, connect a resistor of 1,000 ohms or less across the millivoltmeter terminals.) Now turn the potentiometer on the oscillator gently, looking at the output of the AC millivoltmeter or oscilloscope for an increase in signal. You should get a clear and sharp resonance (at around 40 kHz if you check on the oscilloscope or with a frequency meter). You may find that the signals that you can see on your multimeter, AC millivoltmeter, or oscilloscope are rather too small to be easily measured. If so, you may need to build the simple one-transistor amplifier shown. The operating point of the amplifier is stabilized by negative feedback via a large-value, 3.3-Mohm resistor. Check that the DC voltage at the output is somewhere

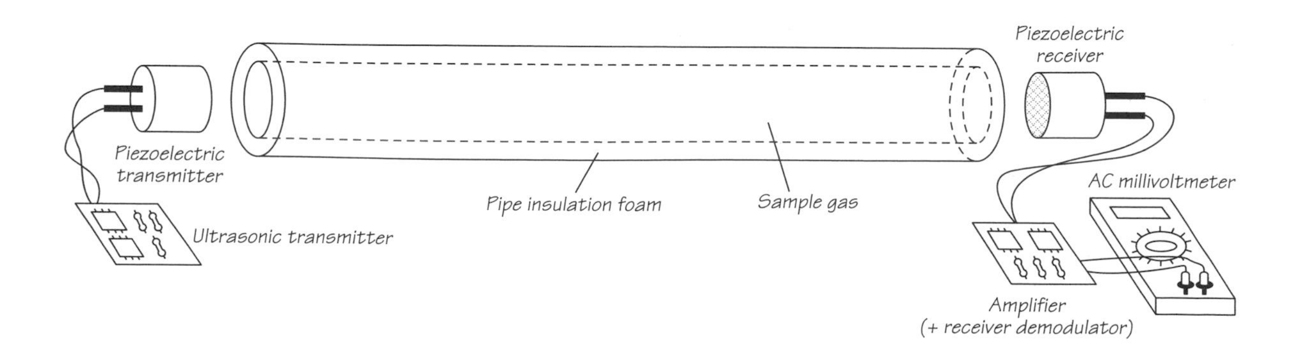

Two-transistor oscillator for ultrasonic gas absorption

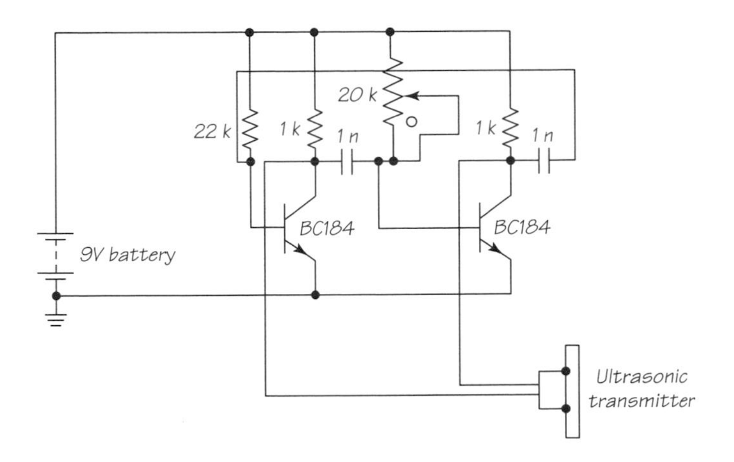

between 2 and 7 volts: this is the range in which the amplifier operating point should fall and in which it will work nicely.

Now you are ready to test some gases. Plug both transducers into opposite ends of a long (approximately 1 m) section of spongy pipe insulation. This material will not transmit any ultrasonic sound and will to some extent absorb sound that falls on it. This means that sound waves have to travel through the gas in a relatively straight line down the tubing. With the transmitter tuned to resonance, you should be able to clearly see a signal traveling down the tube. Now pull out the transducers, squirt in a sample gas, and then immediately plug the transducers back in. Alternatively, you can fill a beaker with a heavy gas and then pour gas from the beaker—much as you might water—to simply flow into the pipe insulation tube, rather than squirting it from a can.

WARNING

Butane is flammable and can form explosive mixtures with air. Keep away from any sources of ignition, and work outside with a plastic beaker. The easiest way to dispense butane is to block up the air holes of a small handheld plumber's burner with tape. Use your nose to detect the butane; if you can smell it, then there may be a danger. Most commercial butane is "stenched," that is, it contains a powerful-smelling additive. You can smell it by virtue of the parts per million or two of a smelly vapor such as ethyl mercaptan, C_2H_5SH. Such additives can be smelled in air well below the concentration at which a butane/air mixture can ignite, which is an important factor in reducing accidents.

Ultrasonic amplifier

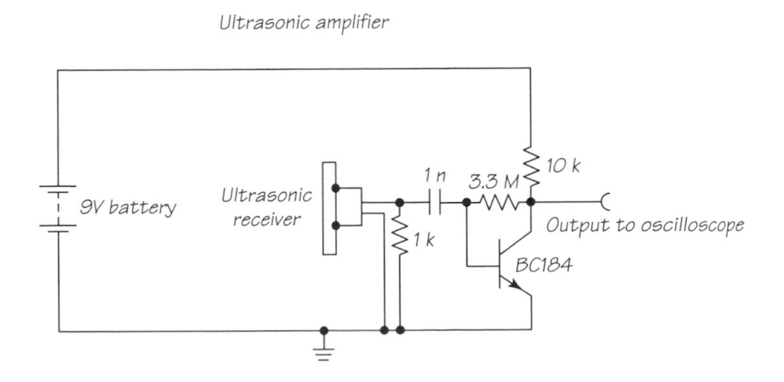

You should see an immediate decrease in the transmitted signal. With a longer squirt of HFC 134a, for example, you should see a complete extinguishing of the signal. Now pull out the transducers and flush out some of the gas with air. The signal should return. You should also see some other effects. After a while, for example, the gas may be absorbed to some extent by the spongy plastic, so a natural loss of gas occurs, with some small recovery of the signal. Try the different gases suggested to see how they affect the transmitted signal.

THE SCIENCE AND THE MATH

Sound is transmitted through gas as a compressive wave. One region of gas is compressed next to the source, and this region then exerts a pressure on the next region, compressing it until it in its turn begins to compress the next region, and so forth.

Wave motion in the gas can be described mathematically by the wave equation for the excess pressure P. Excess pressure P is the pressure relative to the atmospheric pressure P_a prevailing in the gas, in other words, pressure in gas = $P_a + P$. The wave equation is

$$\partial^2 P/\partial t^2 = c^2 \partial^2 P/\partial x^2$$

where c is the speed of sound and t time. The solutions to the equation are traveling waves such as

$$P = P_o \sin(\omega t + kx),$$

which is a wave of frequency $f = 2\pi\omega$ moving along at speed $c = k/\omega$.

Triatomic and polyatomic gases are much more effective at absorbing sound than monatomic or diatomic gases. It is easy to understand why little absorption is possible by monatomic gases: they are close to an ideal gas in the kinetic theory sense, and it is hardly a surprise to find that they are also ideal regarding sound transmission. A monatomic gas can only store energy in the translational kinetic energy of its atoms—the assumption made by the simple kinetic theory of gases. So, at least approximately speaking, no energy can disappear from the sound wave in a monatomic gas.

With diatomic and polyatomic gases, however, energy can be stored in the rotation of the molecules and also, more subtly, in the vibration of the atoms in those molecules with respect to each other. This energy can be stolen from the passing sound wave, which is therefore absorbed.

The compression and expansion process of a monatomic gas stores a little energy in the form of heat. This takes place by adiabatic heating: the gas is hotter in the compressed state and cools as it expands. The energy stored in the random motion of the molecules is returned afterward almost instantly, and with almost perfect efficiency. The result is that this loss of energy is negligible at ordinary sound frequencies. For all sound frequencies up to a megahertz, there are no mechanisms for appreciable energy loss. At higher frequencies, heat production grows, and the absorption coefficient of monatomic gases grows with frequency f as f^2.

With polyatomic molecules, some of the energy *can* be stored for a finite time: rotations of molecules, once excited, don't de-excite instantly. With vibrations and combinations of vibration and rotation, the de-excitation process may be even more difficult, and much of the energy released on de-excitation will be released as heat. Energy from a passing wave that goes into heating up the polyatomic gas will be retained by that gas for microseconds or longer, so waves of that period and shorter will be highly absorbed. The simple f^2 absorption growth does not apply, and very high coefficients of absorption can be seen.

The essential difference between diatomic on one hand and tri- and polyatomic gases on the other hand is not so obvious. In fact, in a diatomic molecule, because of quantum effects, the minimum vibration energy that can be stored is much larger than the thermal energy. The minimum vibration energy is several eV, whereas room-temperature kinetic energies reach the order of 1/40 eV. With tri- and polyatomic gases, however, this lowest energy of vibration is much lower, so these modes of energy storage are more easily excited.

The excitation of rotations and vibrations is a probabilistic process involving a small chance within many excitation collisions: the overall relaxation time is the multiple of many collisions times a small probability. The relaxation time of a molecule is thus inversely proportional to pressure for quite wide changes in pressure, since the collision frequency of molecules in a gas is proportional to pressure. This is a curious result. If you plot the absorption peak of a gas versus (frequency × pressure) on the x-axis, you get the same peak, no matter what pressure or frequency you choose.

The complete story of vibration and rotation excitation in gases is complex, one that tells a fascinating story about the chemistry and physics of molecules. The simple measurement of ultrasonic absorption at different frequencies can be used to unravel some of these mysteries—as A. B. Bhatia shows in his book *Ultrasonic Absorption.*

And Finally . . . Spongy Liquids

What other gases can you get to measure absorption? What about exploring the effects of high–vapor pressure liquids too? (But be aware of the dangers of using flammable liquids with electrical equipment, even battery-powered equipment.)

You may like to try passing modulated ultrasound through gases. With this approach, you can hear an audible signal—the modulation on the ultrasonic wave—attenuated by the gas. The circuit diagrams show a modulated ultrasonic

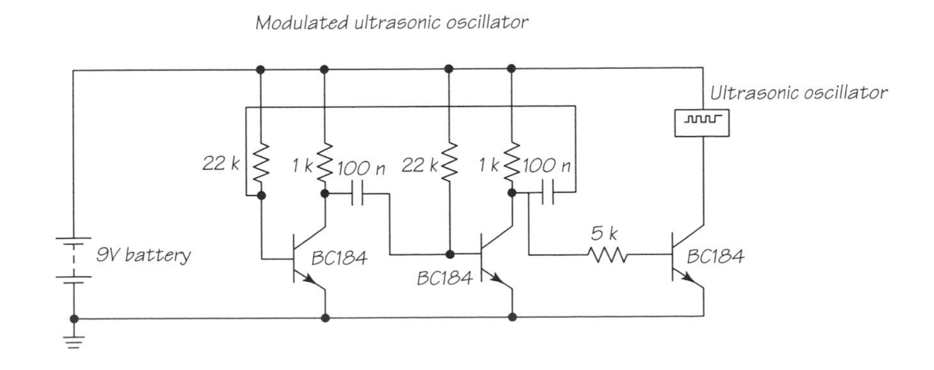

Modulated ultrasound receiver

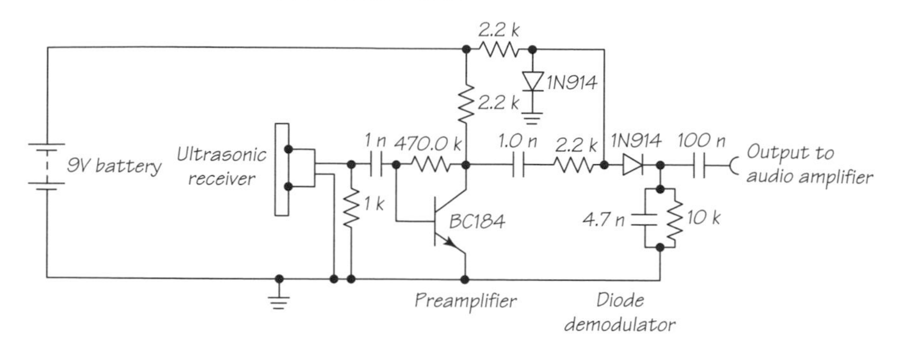

generator for the transmitter and a diode-based demodulator for the receiver. The receiver output is fed to a standard audio amplifier and loudspeaker. The modulated ultrasound method gives a very direct feel for ultrasound absorption.

A similar direct feel for the absorption of gases can be obtained with a simple oscillator transmitter, using a superheterodyne approach to shift the received signal to an audible frequency. A circuit for this approach was diagrammed in the "Bat Doppler" segment of one of my earlier books, *Vacuum Bazookas, Electric Rainbow Jelly.* Commercially produced bat detectors are available, intended to pick up the echolocation signals of pipistrelle and similar bats, that could be used in this way.

REFERENCE

Bhatia, A. B. *Ultrasonic Absorption: An Introduction to the Theory of Sound Absorption and Dispersion in Gases, Liquids, and Solids.* Oxford: Clarendon Press, 1967.

31 Soda Water Gas Analyzer

Science is really the search for simplicity.

—Claude A. Villee

If the basic idea is too complicated to fit on a T-shirt, it's probably wrong.

—Leon Lederman

As you read this, you may be enjoying a glass of soda water of some kind. Since its invention by Joseph Priestley in the 1780s, soda water has become immensely popular. There must be few people who have never tasted one of its many variants, and making and selling sodas is one of the world's biggest businesses. Coca-Cola is reckoned to be the most valuable single commercial brand.

Back in the pioneer days of science, the power of human senses was invariably employed to analyze new chemical compounds. There were few relevant instruments, and even those were poor in quality. So the eye and the nose were frequently used in what we would now consider hazardous ways. All gases except mixtures of an inert gas with oxygen containing at least 20 percent of O_2 are either poisonous or asphyxiant. Many gases are so lethal that they are subject to rigorous controls to ensure they never contact human lungs. Nevertheless, the pioneers of chemistry had no option but to sniff any new gas.

Priestley sniffed carbon dioxide and noted the acid but not too unpleasant effect. He may well have come to dissolve CO_2 in water and then drink the resulting soda water as a result of using a pneumatic trough. The pneumatic trough was a large bowl full of water with a gas pipe entering at the bottom, which he used to fill a gas jar via downward displacement of water as gas bubbled into it. The water in the trough would have become saturated with CO_2 after a while.

Here we explore an unusual property of simple CO_2/H_2O soda water: it has an electrical conductivity greater than that of plain water, an increased conductivity roughly proportional to the amount of gas added.

What You Need

- ❑ Small syringe (5 cm^3)
- ❑ Large syringe (20 cm^3 or more)
- ❑ Air pump (e.g., fish tank air pump)
- ❑ Metal plates
- ❑ Connecting wires
- ❑ Flexible plastic tubing, 6 mm (1/4 inch) in diameter
- ❑ Deionized (DI) water
- ❑ CO_2 source (human breath is a source of 5 percent of the CO_2 in air)
- ❑ Oscillator with transistors, diodes, capacitors, and resistors (see diagram)
- ❑ Meter (e.g., 0–100 μA)
- ❑ Glue

What You Do

Assemble the apparatus as shown. The U-tube connects the syringes so that the small syringe, which contains about 1 cm^3 of water, will not run out of water from evaporation; the larger syringe serves merely as a reservoir. The circuit shown is a symmetrical "multivibrator" oscillator, which produces a square waveform with opposite phases on the two connection points. The meter is connected via the diodes to measure the current through the cell.

With pure deionized water, the cell's conductivity should be small, and little current will flow. Now blow a gentle bubbling stream (two or three bubbles per second or less) through the syringe. Within 30 seconds or so, the current on the meter will increase. The CO_2 in your breath has dissolved in the water and formed carbonic acid. This is a weak acid but, like all acids in aqueous solution, has a certain degree of electrical conductivity, thanks to the charged ions form-

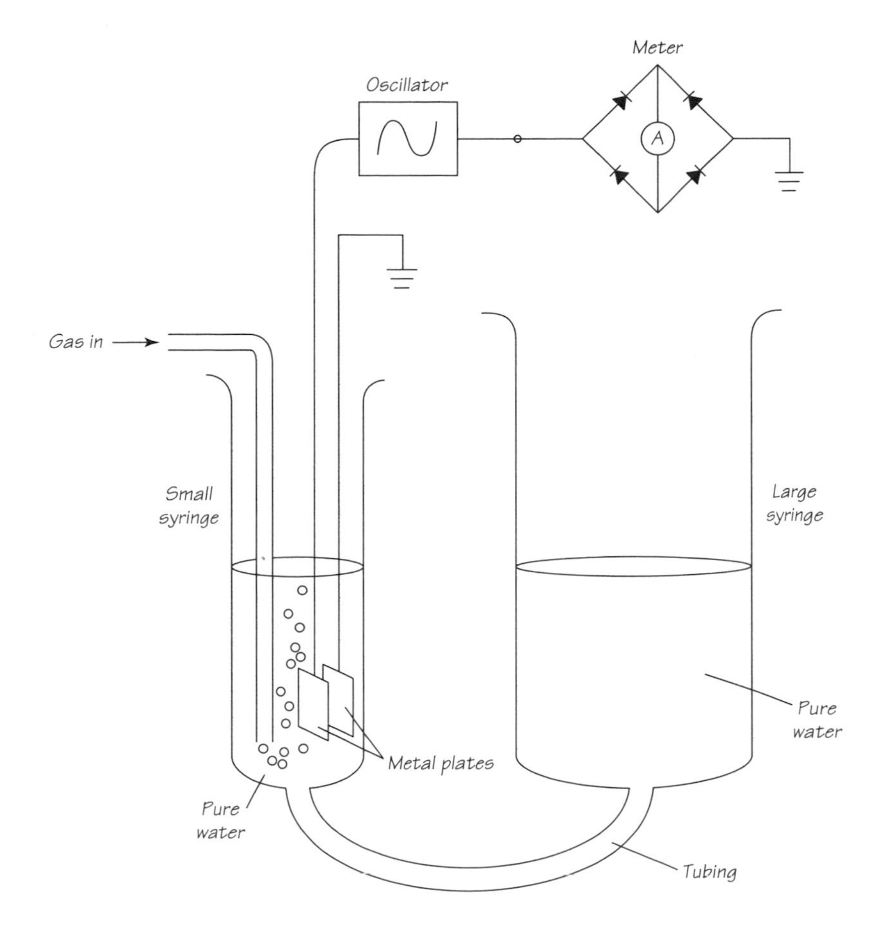

ing in the solution. Now discontinue and blow pure air through the device—but not air from your lungs! Use the air pump. On a longer timescale—about 5 minutes—the conductivity will fall as the pure air washes out the carbon dioxide, a process sometimes known as "sparging." The plot shows the kind of expo-

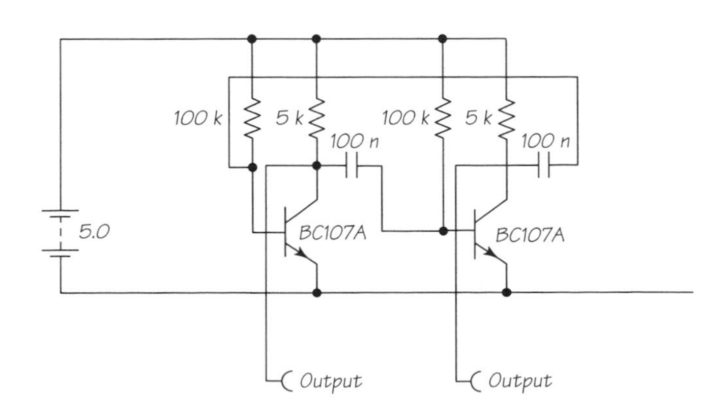

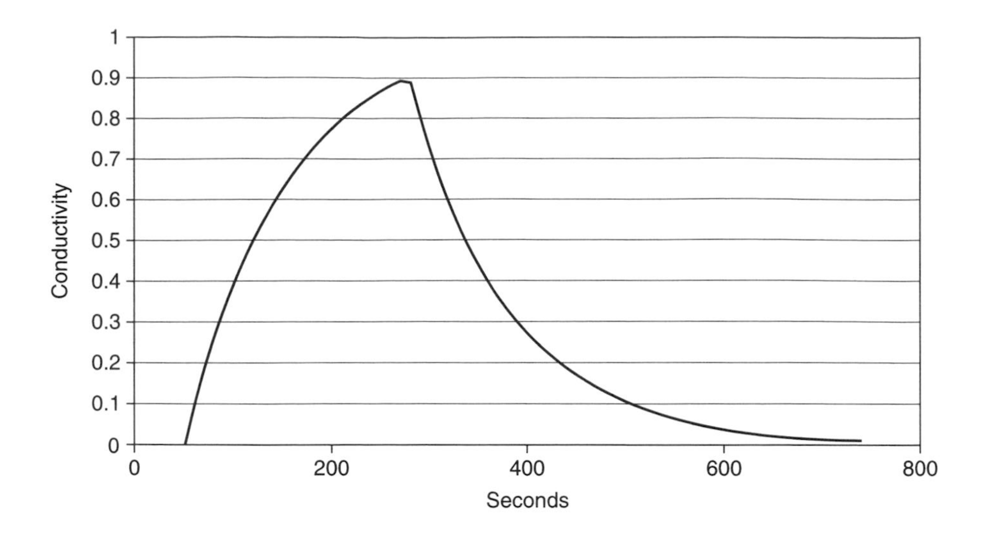

nential changes in conductivity that the analyzer should register when gas is added and then when it is flushed out by air.

This may well be the world's simplest gas sensor! Try using a pure CO_2 source from a soda drink–making machine like a Sodastream or a soda siphon. It is tricky releasing a steady stream of gas, but it is possible. I converted my Sodastream by adding a small piece of small-bore plastic tubing to the carbonating probe and thereby leading the gas out of the machine to the syringe. Press very gently on the operating lever and you can, with care, release a slow flow of CO_2 that will quickly saturate the water, yielding a large conductivity change.

THE SCIENCE AND THE MATH

When carbon dioxide dissolves in water, the gas molecules mostly bump around among the water molecules. However, a small proportion of the gas molecules reacts with the water to form hydrogen carbonate (HCO_3^-) ions as follows:

$$2H_2O + CO_2 \rightarrow H_3O^+ + HCO_3^-.$$

We use a small AC current, rather than DC, because we don't want to electrolyze the solution. This would probably result in the metal of the electrodes being ionized, leading to a permanent increase in conductivity. In other words, with iron electrodes,

$$Fe \rightarrow Fe_3^+ + 3\ e^-.$$

The rate of washout depends primarily upon the rate of dissolution of CO_2 into the water from the gas in the bubbles, and the rate of outgassing of CO_2 from the liquid into the gas in the bubbles. Inside each bubble, there will be an equilibrium between the CO_2 molecules inside the gas phase of the bubble and the CO_2 molecules in the bubble's water-based walls. The rate Q_{sp} at which CO_2 leaves the bubble's walls will be proportional to the area of the walls and the concentration x of molecules in the walls. The rate Q_{dis} at which gas leaves the bubble will be proportional to the area and partial pressure in the bubble. This dynamic might be written as yP, where y is the concentration and P the total pressure:

$$Q_{sp} = K_{sp}Ax$$

$$Q_{dis} = K_{dis}AyP,$$

where A is the area of the bubble surface and the Ks are constants. Thus the total rate of dissolution/sparging will be $Q_{dis} - Q_{sp} = AK_{dis}yP - K_{sp}x$. The system reaches equilibrium when this calculation produces zero, in other words, when

$$K_{sp}x = K_{dis}yP$$

but $x = km/V$ where m is the mass dissolved in the water, V is the volume of water, and k is a constant.

$$Q_{dis} = dm/dt.$$

So $dx/dt = k/V\ dm/dt = k/V\ (Q_{dis} - Q_{sp})$

$$dx/dt = k/V\ (K_{dis}AyP - AK_{sp}x) \text{ and}$$

$$dx/dt = kA/V\ (K_{dis}yP - K_{sp}x),$$

where K_{dis}, y, and P are constants. This equation can now be integrated to express the change of concentration of x with time:

$$x = (K_{dis}/K_{sp})yP(1 - e^{-t/\tau}).$$

This constitutes a curve heading from zero to an asymptotic value of $x = (K_{dis}/K_{sp})yP$.

A similar curve from the settled output back to zero is followed by the analyzer when it is sparged with pure air. As required for a gas analyzer, the settled output of the conductivity, which is proportional to x, is proportional to the gas concentration y in the gas being measured.

A few other features are worth noting. The signal can be increased by increasing the pressure P in the conductivity cell. This increase could be generated by a pump feed, perhaps regulating the outlet flow using a simple spring-loaded check valve or a true back-pressure regulator of some kind. The time constant τ is given by

$$\tau = V/(kAK_{sp}).$$

For an analyzer, a short time constant is usually an advantage: τ can be reduced by decreasing the volume of water V or by increasing the contacting bubble area A.

Dissolving gases in liquids, and the reverse process—sparging—of removing gases from liquids, are both important industrial processes. Thus the design of gas and liquid contactors such as bubble spargers and packed-bed columns is crucial in chemical engineering. The rates of dissolution or sparging are often limited by surface films in the gas and the liquid, and equipment design must therefore reflect that. Although dissolved gas may circulate apparently freely throughout the bulk liquid, the surface film—constituting the last few microns of the liquid—may be depleted of dissolved gas as that gas escapes into the gas phase, which may greatly reduce the rate of dissolution.

And Finally . . . Sand and Beads

Surface films may well inhibit the speed of dissolution and sparging of CO_2 in your soda water gas analyzer. So there is probably scope for greatly improving the speed of the analyzer response. What about, for example, including an inert, finely divided solid such as carefully washed sand in the analyzer tube? You could try a small amount of very fine solids in suspension. Or you could use coarse solids, such as small glass beads, added in excess to the water. But will these measures speed up the exchange of gases into and out of the liquid?

What about the effect of dissolved materials that mop up CO_2? Would they reduce the conductivity at first and then increase it as they release CO_2 during the washout phase?

Clearly, you have to be careful in choosing your compounds. Provided you use a compound that does not, of itself, result in an increase in conductivity, the analyzer should still work. Common CO_2 absorbers like lime ($Ca(OH)_2$), however, are ionic in solution. Perhaps a physical absorber such as acetone could be added? What effect would that have on conductivity?

32 Electric Plastic

> I love Los Angeles. I love Hollywood. They're beautiful. Everybody's plastic, but I love plastic. I want to be plastic.
>
> —Andy Warhol

The basic tool of quite a few people in industry and commerce, as well as countless university investigators, is a sensor that detects the chemical properties of a substance or mixture, providing an electrical output. Walk into any analytical laboratory or any chemical process factory, and you will see a battery of instruments, everything from mass spectrometers and gas chromatographs to oxygen fuel cells and moisture meters. Large, small, complicated, or simple, at the end of the day all that the vast majority of these instruments do is to convert a chemical signal into an electrical one, a signal that can be read into a computer where the data can be processed.

A mass spectrometer, for example, uses multiple vacuum pumps to remove all the air from a chamber. It then sends an electron beam through a sample chemical or gas to ionize it. The spectrometer accelerates these ions through a magnetic field, which separates ions of different mass. Finally, it catches the ions and counts them with a multiplier or other sensitive current detector. But the overall scheme is still chemical in, electrical signal out. (It is virtually the same in living creatures; our chemical senses—smell and taste—provide an electric signal to our nerves, and hence to our brains, in response to chemical stimulus.) One of the key goals of industrial sensor development is to take a chemical input and convert it, using as little equipment as possible, into an electrical output.

You do not need to be a carbon-based life-form to appreciate that carbon is an extraordinary element. It forms diamond, the hardest material known, and can be assembled into minute geometric footballs (the buckminsterfullerenes) or into the greasy solid (graphite) used in pencils. It has one of the highest melting points known and was once used for the filament of early lightbulbs. It is arguably the most extraordinary element in chemistry, for it forms more chemical compounds than all the rest of the elements put together.

Electrical conduction is yet another extraordinary property of carbon, one we don't normally appreciate because many elements are metals that have this property too. But electrical conductivity is highly unusual among the nonmetals of the periodic table. Sulfur is an excellent insulator, as are boron and phosphorus. (There is a rare allotropic modification of phosphorus known as black, violet, or "metallic" phosphorus that does conduct electricity.) The semimetals like silicon and selenium are also almost insulators in their completely pure form. They are rendered electrically conductive by tiny quantities of certain elements that add or, curiously, remove electrons to or from their lattices of atoms.

Here we investigate the unusual properties of a type of plastic film designed to avoid the generation of static electricity. The film is made by adding large amounts of finely divided carbon to an otherwise insulating plastic. The material is thus a composite, whose electrical conduction relies on the contact of the carbon particles with each other. But the plastic can be modified by absorbing tiny quantities of solvents and oils; when the plastic changes, so does the conductivity of the composite. The result is that Holy Grail of the sensor industry: a solid-state chemical sensor with a convenient electrical output, so cheap that you can hardly measure its cost at all.

What You Need

- ❑ Piece of black conductive plastic
- ❑ Multimeter
- ❑ Clothespins
- ❑ Sample liquids such as nail varnish remover (acetone), rubbing alcohol, paraffin, and auto paint solvent (xylene)

And, if testing slow-reacting liquids like oils

- ❑ Data logger

What You Do

This device is child's play to make: it is essentially a piece of plastic bridging the two probes of a multimeter set to a resistance scale. The plastic is slightly special: it must be electrically conductive, which is achieved by adding carbon black to a nonconducting polymer. I used the black conductive plastic made for packaging integrated circuits, which need protection from static electricity. The type I chose was about 75 microns thick, is conductive throughout its bulk, not just on a surface skin, and is a PE material (polyethene, more commonly called polyethylene). If you can't find any of this plastic around your house or school, ask at a store that keeps electronic parts or at a computer repair shop, or ask some people who build or repair their own computers or electronics.

First you should clean off any grease or paint from the plastic, using a solvent like acetone or rubbing alcohol if possible. This removes the less conductive lubricants and surface coatings such as paint marking. Cut a strip of the black conductive plastic and form a narrow neck region just a few millimeters in width and 10 mm long, so that a 5- or 10-mm-diameter spot completely bridges the plastic, ensuring that current will have to flow at least 5 or 10 mm across the sample spot. Now fold the strip twice around each probe and clip it on firmly with a clothespin, squeezing the jaws of the clothespin to ensure that it grips the plastic firmly across the full width of the strip. Switch on the multimeter to the resistance range, and your electric plastic sensor is ready to use!

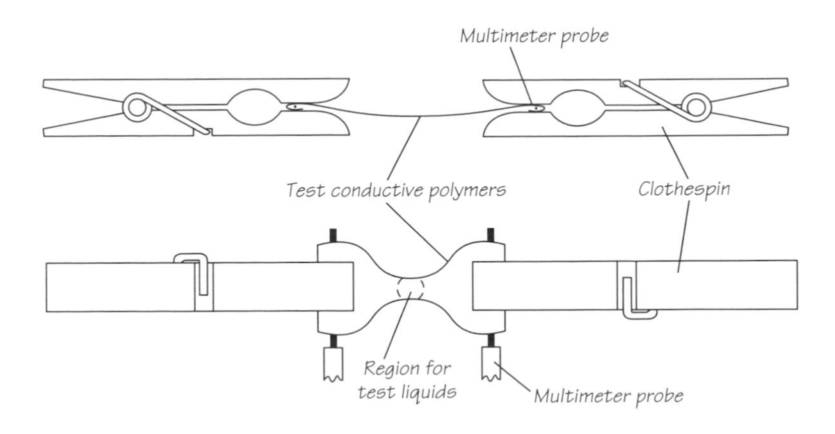

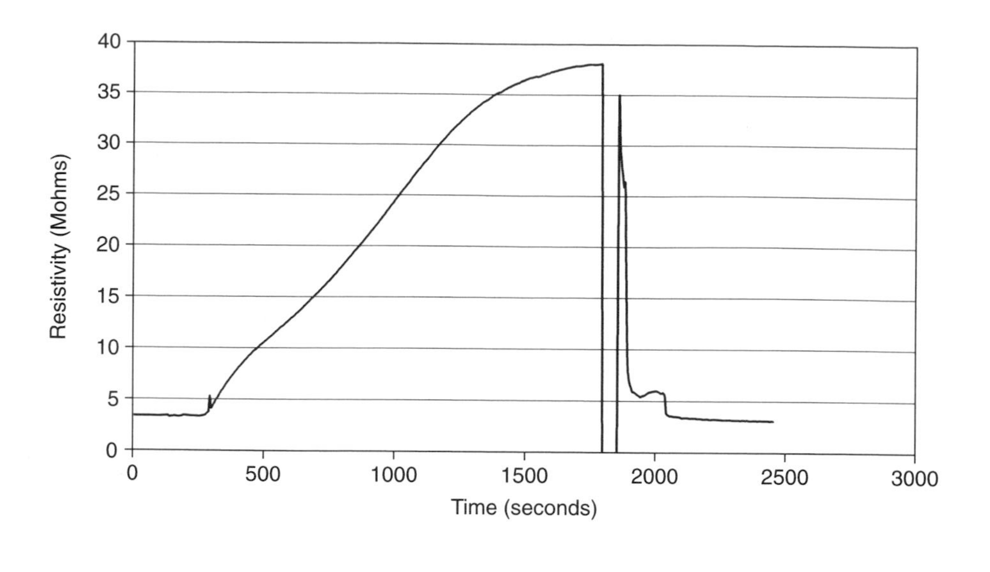

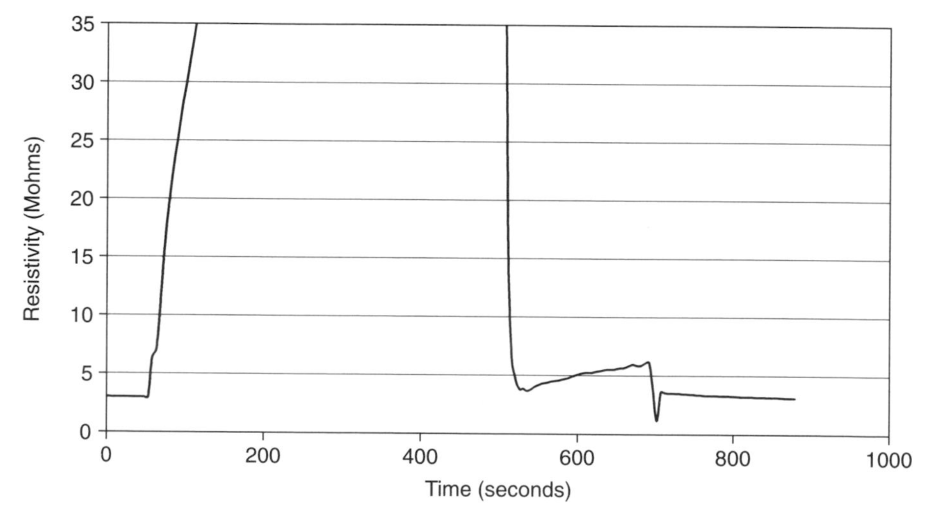

You may notice a transient change of resistance after preparation, but it should settle down to a steady-state value between about 0.5 and 5 Mohms. This is your zero value. Now add a sample of a polar solvent like water, and note the slight change of reading. A more powerful solvent such as acetone will produce a bigger change. Almost always, it will be a change to a higher resistance value. Now try a xylene-based solvent like auto paint thinners. You should see a massive increase in the resistance, probably going off-scale on your meter in less than a minute. You will need to remove the visible solvent and then blow, perhaps warming the tape gently with a reading lamp placed a few inches away, to remove the solvent, before seeing the resistance zoom back down.

A sample of a light lubricating oil like cycle oil should produce a much smaller change. A paraffin or white spirit product will yield an intermediate response, perhaps nearly as large as that of xylene. These oily compounds take longer to respond than xylene, something in the region of 10 minutes. If you are going to test oils and other materials that take a longer time to react, it is best to hook up a data logger. The first graph shown plots the responses I had from one particular electric plastic sensor, when I started my data logger, applied a sample spot of paraffin, and then waited for a full or off-scale response, followed by some cleanup with acetone to restore the sensor to its original resistance. The second plot shows a xylene test.

How It Works

What we see here, chemically speaking, is that like goes with like. A nonpolar long-chain hydrocarbon solvent tends to be attracted by, and wriggle its way into, a similar long-chain hydrocarbon polymer. Once swollen, the polymer conducts less well, which raises the resistance.

Other solvents—highly polar ones such as water, for example—won't be attracted at all into the polymer. They won't even wet the electric plastic surface. Less polar solvents such as acetone will wet the surface and may penetrate a little, but they won't do much either. This result is surprising to anyone who has seen how acetone-based mixtures like nail-polish remover can dissolve polymers like the varnish or paint on wood furniture.

Solvents will sometimes dissolve polymers. Many polymers, however, like the PE in my sensor, do not dissolve but are nevertheless affected by solvents. In the case of PE, the solvent penetrates between the long chains of the polymer molecules, just as solvent penetrates between the molecules of the solid that it is dissolving. However, the long chains of the PE will not float around freely in the solvent, which would be dissolution. They are still somewhat held together, either by cross-links between the polymer chains or because those chains are exceedingly long and tangled together. Once a solvent has penetrated the polymer, it swells slightly. In itself, this reaction would not be enough to notice. That slight swelling, however, pushes the conductive carbon particles in the polymer apart slightly, pushing many of them out of contact with each other. Now the solvent itself, and the PE polymer, do not conduct. If the carbon particles are not touching each other, they will not conduct either. The net effect is that a slight swelling of the electric plastic leads to a big decrease in conductivity.

THE SCIENCE AND THE MATH

The theory of solvent-polymer interactions has been carefully studied in chemistry because of its importance to the preparation of polymers and to some of their applications. These include, for example, solvent-based paints and adhesives; swelling of rubber seals that are contaminated by solvents; or the use of nonvolatile solvents as "plasticizers" to make normally brittle polymers flexible.

A number called the solubility parameter, with symbol δ, can be defined for solvents. This is calculated according to:

$$\delta = \sqrt{([\Delta H - RT_V]/V)},$$

where ΔH is the heat of vaporization, R is the gas constant, T_V is the vaporization temperature, and V is the molar volume at vaporization. The parameter δ can be conceived of as the square root of the cohesive energy density, which measures how much attraction the solvent molecules have for each other. Polar solvents have high values of δ—the molecules of solvents such as water or alcohol have a strong separation of electric charge and therefore attract each other strongly. Nonpolar solvents such as paraffins have low values of δ; their molecules have little separation of electric charge and so attract each other weakly.

If you mix solvents of similar δ, they will be fully miscible—they form an intimate mixture at the molecular level, physically indistinguishable from a pure liquid. This is because it is energetically favorable for them to form a mixture. In moving from pure solvent to mixture, the molecules hardly know that their neighbors have been replaced, because those neighbors are so similar. And entropy will then drive the dissolution process forward. Entropy is the natural tendency, driven by the statistics of many random events, toward disorder, which thus favors the formation of mixtures.

If you mix solvents of significantly different values of δ, however, they will not form stable mixtures. Instead, they separate into globules or, if they are of different densities, into layers. It is not energetically favorable for them to mix intimately: you have to supply energy to break the electrostatic forces that bind neighboring molecules together in the pure liquid. You would not get energy back from the attractive force of those neighbors with a different molecule in the mixture.

Adding a solvent to a polymer is similar to mixing solvents. If you add a solvent whose δ-value is the same as the δ of the polymer, then it will swell and may, if it has a low molecular weight and is not cross-linked, dissolve to form a solution in the solvent. So when you add a paraffin to polyethylene—which is in effect a high-molecular-weight paraffin—the δ-values are similar, and thus the solvent penetrates and swells the polymer. But if you add ethanol to polyethylene, the δ-values are widely different, and there is virtually no swelling. Here are some δ-values:

Polymers	δ	**Solvents**	δ
PE	7.88		
Polyisobutylene	8.06		
Natural rubber	~8–8.3	CCl_4	8.6
Polystyrene	8.56	Toluene	8.9
Polyvinyl acetate	9.38		
Monsanto N5400 Vyram rubber	10.3		
Nylon 66	13.6	Ethanol	12.7

A caveat must be sounded. Some polymers and solvents will have much stronger interactions. Some solvents may break cross-links in the polymer or even attach the chains themselves, or they may form strong adduct bonds. So the solubility parameter is not the whole story.

And Finally . . . Other Polymers

Why not try other high-conductivity polymers? These should ideally be bulk conduction polymers such as the material we tried above, again filled with conductive carbon particles throughout, or perhaps thicker antistatic packaging films. There are also polymer films with conductive surface coatings. Although these might work, the conductive coating is thin and easily damaged or dissolved, so they are likely to be less suitable.

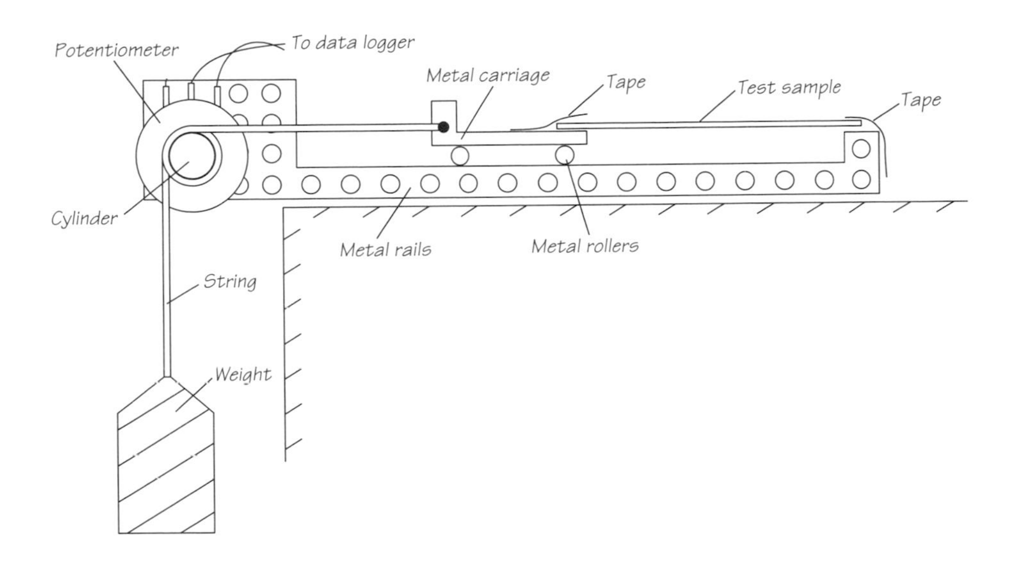

You don't have to use conductive polymers if you use a mechanical sensing of their reaction with chemicals. The diagram shows an arrangement that senses small changes in length of a polymer and generates an electric output via the rotation of a potentiometer (variable resistor).

REFERENCES

Griskey, Richard G. *Polymer Process Engineering.* New York: Chapman and Hall, 1995.

Hall, Christopher. *Polymer Materials: An Introduction for Technologists and Scientists.* 2nd ed. New York: Wiley, 1989.

Chemical Corner

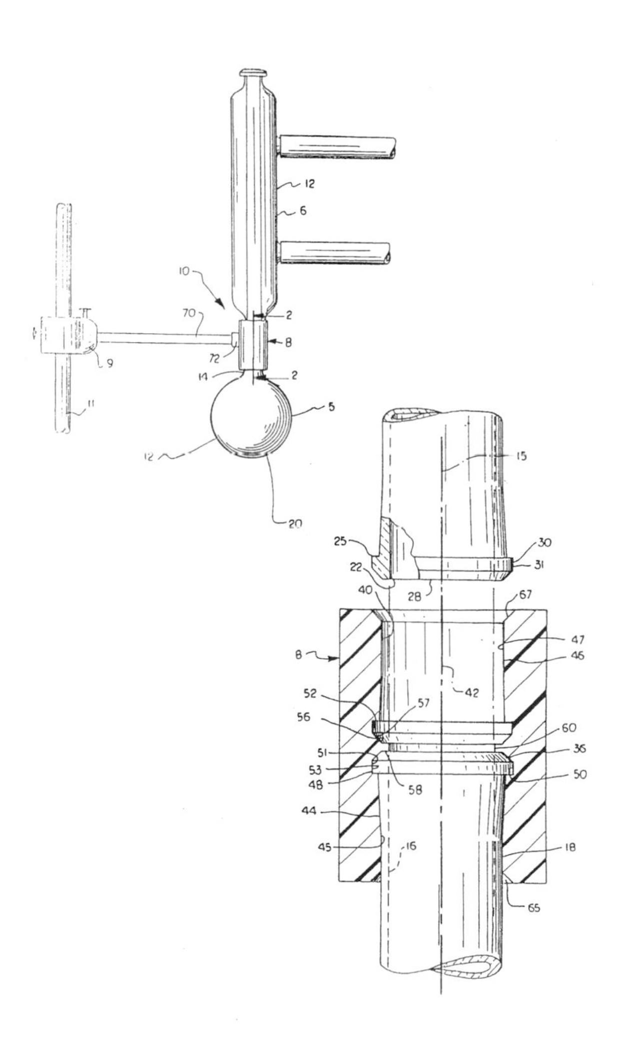

For me chemistry represented an indefinite cloud of future potentialities which enveloped my life to come in black volutes torn by fiery flashes, like those which had hidden Mount Sinai. Like Moses, from that cloud I expected my law, the principle of order in me, around me, and in the world. . . . I would watch the buds swell in spring, the mica glint in the granite, my own hands, and I would say to myself: "I will understand this, too, I will understand everything."

—Primo Levi, "Hydrogen," *The Periodic Table*

In this last section of the book, we turn our attention to a pair of curiosities that have hitherto lain buried in chemistry's corner cobwebs. If chemistry is defined as the behavior of the outer electrons of atoms, then our first project involves a kind of chemistry, since it involves the removal of electrons from atoms. The electrons are removed in this case not by any normal chemical process, however, but by a mechanical process—literally rubbing them off. Our last project is a bit of "chemical magic" with the element iodine, which I have turned around to effect a peculiar sort of printing.

33 Electric Sand

The great sound . . . in some remote places startles the silence of the desert. Native tales have woven it into fantasy; sometimes it is the song of sirens who lure travellers to a waterless doom; sometimes it is said to come upwards from bells still tolling underground in a sand-engulfed monastery; or maybe it is merely the anger of the jinn! But the legends, as collected by the late Lord Curzon, are hardly more astonishing than the thing itself. I have heard it in south-western Egypt 300 miles from the nearest habitation. On two occasions it happened on a still night, suddenly—a vibrant booming so loud that I had to shout to be heard by my companion. Soon other sources, set going by the disturbance, joined their music to the first, with so close a note that a slow beat was clearly recognized. This weird chorus went on for more than five minutes continuously before silence returned and the ground ceased to tremble.

—Ralph A. Bagnold, *The Physics of Blown Sand and Desert Dunes**

*Ralph Bagnold put his deep knowledge of the behavior of desert sand to good use in helping defeat Field Marshal Erwin Rommel's Afrika Korps during the Second World War. He developed a high-speed driving technique for surmounting dunes, and practices such as reducing tire pressure on vehicles when they had to traverse loose sand.

You can split atoms in your own home. Here's how: Rub one solid substance up against a different one and what happens? Minute traces of one substance rub off onto the other. The macroscopic force you apply is enough to break off at least a few atoms of one onto the other. This process is obvious with a piece of chalk and a section of blackboard, but something similar happens when you rub any two hard objects together. Rub two diamonds together and a little of each will rub onto the other. You might not rub off many atoms, but you would certainly rub off some of the electrons of those atoms, and so you could say that you were "splitting atoms."*

If at least one of the two solids is an electrical insulator, then in general you scrape off the outside electrons. This is how static electricity was discovered by the ancient Greeks. They noticed that it occurred when the natural hydrocarbon polymer resin we call amber was rubbed with a cloth. The word *electricity* was derived from the Greek name for amber: *elektron.*

Making electricity by rubbing has fallen out of favor now, but for a long time it was the only way to generate any kind of electric power. A whole science called triboelectricity grew up. Materials like sulfur and wool were noted for their performance. More recently, industrial chemistry has provided us with insulating polymers, many of which are spectacular performers in the triboelectric stakes. Everyone is familiar with the way expanded polystyrene can be charged by rubbing it, and how little crumbs of it become charged and stick to everything. Nylon became so famous for its production of electricity that clothing manufacturers were forced to reformulate nylon fiber with antistatic additives to reduce the phenomenon.

In this project, we use the triboelectric properties of small particles to measure the flow of a stream of particles down a chute.

*The phrase "split the atom" came to mean the splitting of an atomic nucleus in the early days of nuclear physics in the 1920s. It takes a great deal of energy—a million electron volts (MeV) or more—and is a difficult process to achieve. But splitting an electron or two from an entire atom, which comprises the nucleus and its orbiting electrons, is much easier, needing only a few electron volts (eV) to achieve. This kind of atom splitting happens in everyday chemical reactions.

What You Need

- ❑ Finely granulated materials, starting with 0.2- to 1-mm-diameter granules (e.g., dry sand like "silver sand" or washed river sand; table salt; plastic molding granules)
- ❑ Section of U-channel, perhaps a piece of plumbing tube cut in half, made if possible from several different polymers (e.g., UPVC, polypropylene, Teflon)
- ❑ Funnel with restricted nozzle (e.g., bored cork)
- ❑ Scoop
- ❑ Bowl to receive flow
- ❑ Thin metal plate (e.g., aluminum litho plate)
- ❑ Multimeter with 200-millivolt range (or data logger and PC)

- ❑ Scales (e.g., digital kitchen scales)
- ❑ Wires
- ❑ Clamps
- ❑ Wooden supports

What You Do

First make sure that your chosen granulated materials are completely dry and will flow freely. You may need to dry the materials in an oven. Salt—even table salt especially prepared to flow freely—may need drying. Sand, as supplied by hardware stores or as shoveled from river banks, will definitely need drying: set

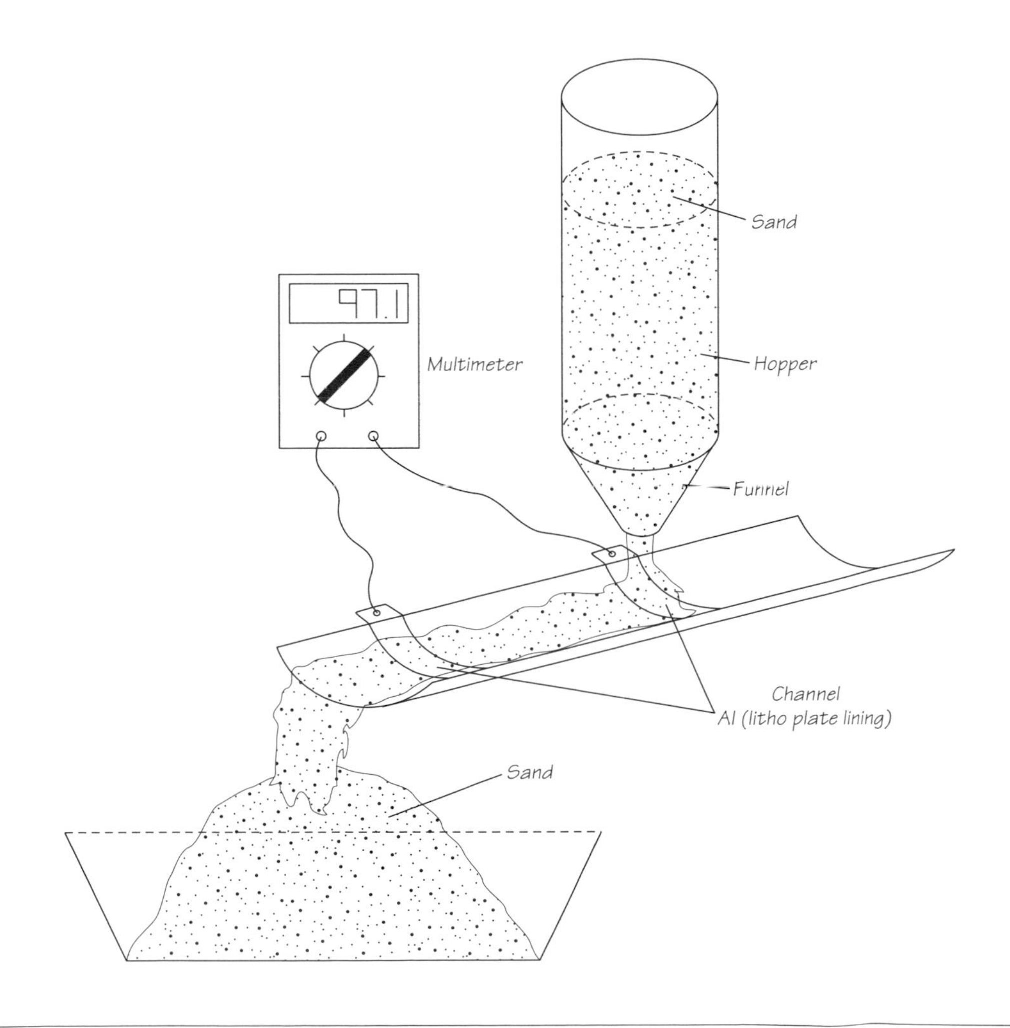

your kitchen oven, preferably the fan-assisted type, on low (180 °F or 100 °C), and put in the sand, layered only 6 to 12 mm (1/4 to 1/2 inch) deep on metal shallow trays. Even after drying, some sand won't flow freely and predictably: sand composed of rounded grains flows much more freely than sand with angular grains. Examine a few grains under a microscope and you will easily see whether your sand has this problem. Obtain another sample of sand if necessary.

You may find that you can get a few handfuls of 3-mm plastic molding granules from a local injection molding shop. Alternatively, try the 1-mm polymer granules used for water softening. Laboratory chemical suppliers can offer many more possible substances to try, in a variety of granule sizes. Finally, don't forget to sieve any random natural materials you use to eliminate oversize grains and interlopers like grass stems or leaf fragments.

Next block the outlet of the funnel to achieve a suitably regulated flow. I found that a few grams per second was a good beginning flow rate. You should aim for a flow rate at which the particles will not pile up on top of each other but instead slide down the chute more or less individually, with each particle touching the chute most of the time. Hold the funnel above a bowl on the scales for about a minute to check the flow rate. You should find that a hole measuring 2 to 5 mm will be a good start with the sand grains and other small particulates.

I made my chutes from a piece of PVC tube used for sink wastewater, cutting it in half to make a U-shaped channel. I then bent two 25-mm strips of litho plate to fit snugly on the inside surface of the U-channel and glued them in place on the chute. I left a piece of the plate sticking out from the side (see diagram) for easy connection of the meter or data logger using alligator clips.

Incline the chute at a fairly steep angle of about 45 degrees and connect the plates to the multimeter, which should be set to the most sensitive scale. As soon as the material starts flowing, you should see a rapid increase in voltage to a plateau value that is relatively constant, unless you interrupt the flow. The voltage will jig up and down a little, due to the random arrival of the granules. Try logging the voltage if you want truly accurate results: after taking the data, you can then average the results in a spreadsheet. The graph shows the voltage induced on two electrodes across a 75-mm (3-inch) PVC drain channel. Clean, dry quartz sand was trickling from a 7.5-mm (5/16-inch) diameter orifice. The two peaks show electrodes 110 mm (4 1/2 inches) and then 40 mm (1 1/2 inches) apart.

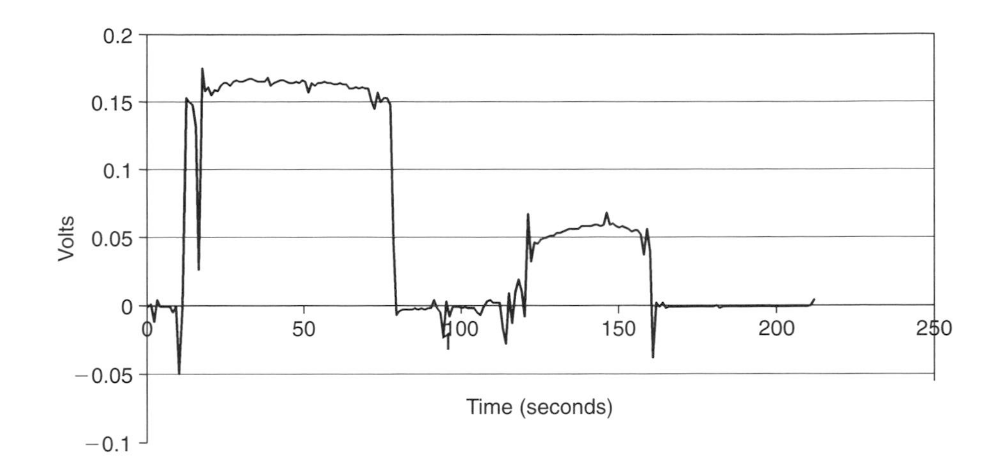

If you change the flow rate, you should see the voltage modulate accordingly. Similarly, if you increase the distance between the chute electrodes, you should see an increase in voltage and vice versa. If you try different materials for the chute, then you will also see a change in voltage. The table provides a clue as to whether you should expect to see a large or a smaller voltage. If you have different sizes of granules available of the same material, then they provide an additional variable you can test.

THE SCIENCE AND THE MATH

The table on the following page represents the "triboelectric series." Although it is rough-and-ready by comparison, it is inspired by the electrochemical series (which gives the much more precise voltages generated by pairs of metals in electrolytes). It provides some idea of which polymers or other substances will charge positively when rubbed against a substance lower in the list. (The lower substance will take an equal and opposite negative charge, of course.)

A Triboelectric Series

Most positive

human skin
asbestos
rabbit fur
acetate
glass
mica
human hair
nylon
wool
fur
lead
silk
Al
paper
cotton
steel
wood
amber
rubber balloon
sealing wax
hard rubber
Mylar
nickel
copper
silver
brass
Au/Pt
sulfur
acetate (rayon)
celluloid
polyester
styrene (Styrofoam)
Orlon
acrylic
saran (polyvinylidene chloride)
polyurethane
polyethene
polypropylene
PVC
KelF
Teflon
silicone rubber

Most negative

Different sources and analysts list varying substances in the triboelectric series. In much of industry, electrostatic charging is a problem that must be minimized, and there are many versions of the series for what is called electrostatic discharge (ESD) control. Ford's *Homemade Lightning* offers more information on the triboelectric series.

Probably the simplest mathematical model for the voltage measured V is that it is proportional to the electrode spacing L and the mass flow rate of particulate Q. In other words,

$$V = kQL,$$

where k is a constant that depends upon the nature of the chute, the particulate, and to some extent on the slope of the chute. Presumably the particles will begin to reach an equilibrium charge as they slide along a longer chute. So we might expect to see an asymptotic form for the voltage as a function of L for larger values of L, that is,

$$V = kQL_o(1 - \exp(-L/L_o)),$$

where L_o is a constant characteristic of the length required for the particles to charge to a maximum voltage. At low values of L, this leads once again to $V = kQL$, but at higher values of L it tends toward an asymptotic value L_o. Is this what you see with your choice of granules and chute material? Similarly, you might expect to see "saturation" effects with large granules, at flow rates at which the particles pile on top of each other, and when they do not rub against the chute at all.

The minimum usable angle to the horizontal depends upon the friction of the particles. If the angle is too shallow, the flow stops and a pile accumulates. This friction is often measured in industry, where it is an important input to the design of hoppers used with granular solids. Industrial labs take a ring and place it on a piece of the proposed hopper material. This constitutes a container, which is then filled with granules. Different downward forces F_n are applied with a piston that fits inside the ring. The shear force F_s required to shift the ring across the surface is measured, and the result is usually expressed as an angle, the wall friction angle Φ:

$$\tan \Phi = F_s/F_n.$$

Values vary between materials, but an initial estimate of 30 degrees is probably a good start. The friction angle Φ determines the acceleration of granules on a chute—which is zero when the chute is at an angle of Φ to the horizontal. Interestingly, Φ also turns up in the formula for the loss of speed when vertically falling particles hit a chute at angle θ to the vertical:

$$V_f = V_i (\cos \theta - \sin \theta \tan \Phi),$$

where V_f is the speed after and V_i the speed before chute impact. V_f will be zero, causing blocking of the chute, if $(\Phi + \theta) > 90$ degrees.

The fluctuations in the voltage are probably mainly due to the statistics of granules. You should expect the number of granules per second N to vary on the order of $\sqrt{N}$. If $N = 100$, then the granules will vary by $\sqrt{100}$ or 10, 10 percent of the mean value. If there are 1,000 granules per second, then the typical fluctuation size will be on the order of $\sqrt{(1{,}000)}$, or 32, which is 3 percent of the average value.

And Finally . . .
Exotic Chutes

Can you find any exotic materials with which to make a U-channel? For example, polystyrene, used by makers of architectural models, can be formed with heat into U-channels. If you can't make a U-channel but can find a tube, then an oblique cut at the top end of the tube will allow you to pour in the granule flow easily. Glass could be used by cutting an angle funnel. Polytetrafluoroethylene (PTFE), commonly known as Teflon, has the capability of higher triboelectric charging than most polymers, but it may be difficult to find in a U-channel form. Teflon tubing is readily available in smaller sizes and can be cut in half. You could also try coating an easily obtained insulating substrate with wax.

Different paints or varnishes are another possibility, and a varnish or adhesive could be used to glue down a fine particulate or a woven material like wool or nylon, although the chute's downward slope would need to be steep to maintain an even flow.

REFERENCE

Ford, R. A. *Homemade Lightning: Creative Experiments in Electricity.* 3rd ed. New York: McGraw-Hill, 2002.

34 Electric Invisible Ink

> The three great elements of modern civilization: gunpowder, printing and the Protestant religion
>
> —Thomas Carlyle, "State of German Literature," *Critical and Miscellaneous Essays,* 1838

However much you might disagree with the other two, most of us today would probably still agree with Carlyle about printing. Printing had been used for quite a while before printed books became available. It was possible in principle but not easy, and impossibly expensive, to print books by carving letters out of wood and then making impressions on paper by using ink on the wood. Presses of both the lever and screw types, which had been used since antiquity in Greece and Rome for squeezing the juice from fruit, were available for the printing process.

Inks that would print well were difficult to produce. Ordinary writing ink, which works quite well with a quill or fountain pen, does not function well in a printing process. (If you don't believe me, try using fountain pen ink for printing.) Printing ink needs to be a little viscous and sticky, so as not to run off the typeface. It must not dry too quickly on the press, even in thin layers, but it must dry reasonably quickly on paper. The ink should also ideally neither dissolve in water nor attack the paper (acidic ink has slowly etched the paper of many ancient manuscripts). In the Middle Ages, printable ink was made using combinations of linseed oil with lampblack (carbon) and other ingredients. More recently, the full power of modern chemistry has produced inks with characteristics carefully tailored to the particular printing process being undertaken, and in a full spectrum of colors.

No matter how clever the ink system, however, it would be useless without further printing inventions. The key invention in printing technology was the movable type press, which enabled Gutenberg and his colleagues to print books in a reasonably economical way. Instead of carving each page onto a wooden plate, they could assemble a page from letters they could use again and again. (The British Library in London exhibits examples of sheets printed from early movable type, including the work of Gutenberg and, astonishingly, sheets of Chinese characters that predate Gutenberg. They remain remarkably clear and readable to this day.) Movable type imposed some problems: the letters had to be of exactly the same thickness. They had to be hammered into place to flatten the printing surface and then inked with a soft pad, so that letters that were slightly too low still received ink. But with all its initial problems, movable-type printing was a revolutionary step forward.

We explore here how an electrolytic chemical process can selectively create colored pigment on a paper surface, exactly in the spot where the printed design requires it. This technique enables us to carry out an unusual kind of printing that does not use conventional ink of any kind.

What You Need

- ❑ Metal sheet (e.g., stainless steel), a little larger than the size of sheets of paper you wish to print on
- ❑ Electrical insulation tape or other self-adhesive plastic, the narrower the better; perhaps precut into shapes
- ❑ Screwdriver with long, narrow blade (which you will be modifying)
- ❑ Metal (e.g., stainless steel) roller, about 40 mm (1 1/2 inches) long, 20 mm (3/4 inches) in diameter, with central hole large enough for screwdriver
- ❑ Retaining collet (e.g., the metal part from a large electric terminal block or "chock block")
- ❑ Paper
- ❑ Starch
- ❑ Water
- ❑ Shallow tank for wetting paper (optional)
- ❑ Cleaning squeegee and cloth
- ❑ Potassium iodide
- ❑ 12-V battery
- ❑ 10A fuse in fuse holder (automotive 12-V type)

- ❑ Wires
- ❑ Alligator clips

What You Do

This printing system relies on insulating parts of a metal plate with a design and then developing ink wherever that insulator is missing by an electrolytic process, using a rolling electrode. The electrolytic reaction we use produces minute amounts of iodine, which reacts with starch to create an intense color.

First prepare the image you wish to print by sticking pieces of the tape or precut self-adhesive shapes onto the metal plate. You need to ensure that the metal plate is clean and free from grease. If you want to discover how fine a resolution the electric-ink printing process can achieve, stick down a test pattern with lines of progressively finer widths. Make sure the tape is carefully pressed down so that there are no bubbles or edges remaining.

Next prepare your ink precursor solution: add a teaspoonful of starch to a cup (150 cm^3) of hot water and bring to a boil, stirring constantly, diluting the increasingly viscous solution carefully to avoid the formation of lumps. Finally, dilute the soupy solution with more water: you should aim for a slightly cloudy watery solution. Make two to three cups (300–450 cm^3) of solution. Stir in a teaspoonful of potassium iodide.

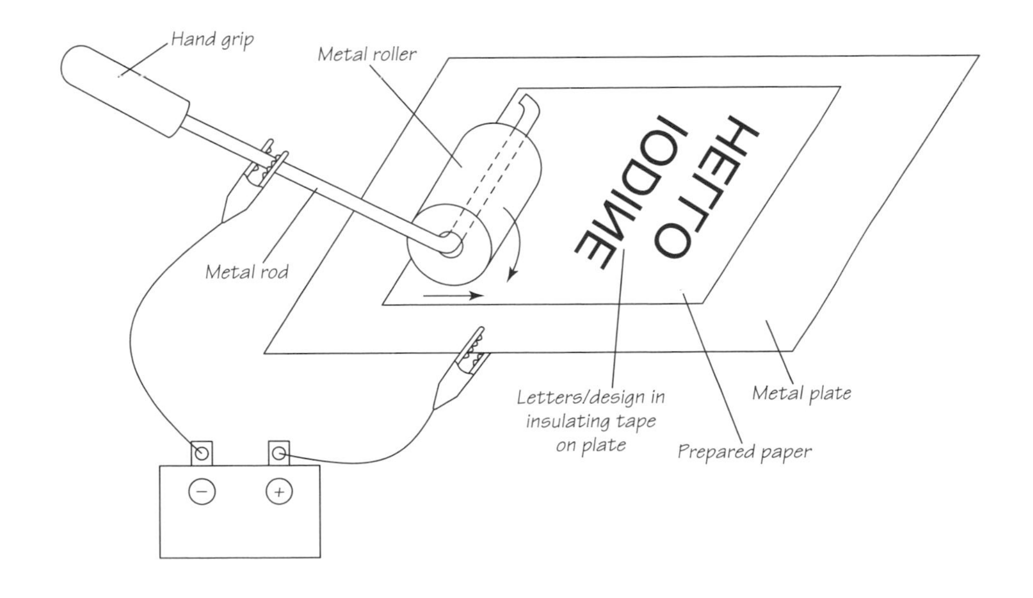

The rolling electrode must now be prepared: clamp the screwdriver in a heavy vise and make a right-angle bend in it as shown in the diagram, fit the roller, and screw the collet to the end of the blade to retain the roller. Attach the negative terminal of the battery to the metal plate with an alligator clip lead, and attach the positive terminal via the fuse holder to the screwdriver shaft of the printing roller. Now put the ink precursor solution into the tank, run a piece of paper through it so that both sides are wetted, and place the paper on the metal plate. Now run the printing roller all over the paper, one strip at a time.

You may think at this point that nothing has happened. But pull the paper off the metal plate and . . . hey presto! You should find that a deep blue pattern, the negative of your design stuck on the metal plate, has been printed on the paper. The deep blue color has been created electrically, by oxidation of iodide ions in solution to iodide molecules; it has only been created where the metal plate contacted the paper, and not where the metal was insulated by the tape.

You will need to experiment a little to find the optimum variables for your system. Try different types of paper, different voltages on the rolling electrode, and different speeds for running the rolling electrode. Try very wet paper versus well-drained paper that is only slightly damp. Try a stronger starch solution or more potassium iodide in the solution. Always clean the metal plate carefully after each "press run."

THE SCIENCE AND THE MATH

What happens in this project is the creation of tiny amounts of elemental iodine through electrolysis:

positive electrode: $2e^- + 2H^+ \rightarrow H_2$

negative electrode: $2I^- \rightarrow 2e^- + I_2$.

This occurs mostly where the negative metal plate contacts the conductive solution, that is, on the underneath of the paper, next to where the metal is bare—but not where the metal has been insulated by the pattern. The iodine then combines with the starch to form a complex compound with an intense blue color:

I_2 (weak brown-purple color) + starch (colorless) $\rightarrow$ Starch-I_2 complex (strong blue color).

The complex is not completely stable. On dry paper the pattern will slowly fade, the deep blue eventually turning to a muddy brown, although the printed design will remain readily visible. Initially, however, the starch/iodine complex is intensely colored. Some of the details of this reaction are still being investigated by chemists, but basically what happens is that iodine (in the form of I_5^- ions) gets bonded by the coils of beta amylose molecules of a soluble starch. The starch lines up the iodine atoms in a regular array down the center line of the amylose coil. Electrons are transferred between the starch and the iodine, which changes the way electrons are confined. This in turn changes the spacing of the energy levels in iodine so that they are just right for absorbing visible light in all but the blue region, causing its blue-black appearance. As the starch molecules dry out, some of this molecular arrangement goes awry, causing the color to fade.

And Finally . . . Alternative Chemistries

Could the printing stability be enhanced in some way? If the instability is due to the loss of water, then the addition of water-retaining material, which might keep the mixture slightly moist, would be useful. Could the use of a different starch be helpful, perhaps a starch that has a natural tendency to retain more moisture?

You could also try using the same principles and equipment with alternative chemistries. I tried using common salt (sodium chloride) instead of the iodide and got some results. With stainless steel electrodes and reversed polarities, I encountered some brown stain—though nothing as powerful as the iodine color—that was due to brown iron salts forming in the paper:

$Fe \rightarrow Fe^{3+}$ (brown hydroxides and complexes) $+ 3e^-$.

With some colored papers, this brown stain seemed more powerful—perhaps a reaction with the dye in the paper to form a complex with the iron ions. On colored paper, too, I found that I could produce a positive copy, in which the design was bleached by the chlorine gas that was produced:

$2Cl^- \rightarrow Cl_2 + 2e^-$.

Chlorine reacts with chromophores, typically centers in organic molecules featuring nitrogen double bonds (as in some synthetic dyes such as methyl orange). It also reacts with conjugated carbon double bonds/single bonds (as in vegetable dyes such as carotene) and then oxidizes the double bonds in the chromophores:

-N=N- or -C=C-C=C- + $Cl_2 \rightarrow$ -NCl-N- or -C-CCl-C=C-, etc.

The positives I got on green fluorescent paper were quite distinct: a pastel shade of green background with the original strong green lettering showing up nicely, although there was some iron staining in places from the stainless steel on the back. The degree to which the paper is bleached depends upon the sensitivity of the particular dye to oxidative bleaching by chlorine: some colored papers could be much more sensitive than those I tried. Maybe you could make your own colored paper with a dye that is particularly easy to bleach—although bear in mind that dye manufacturers try hard to make dyes that are *not* easy to bleach. Food coloring might be easier to bleach, and so a better approach might be to use food coloring to make your own paper.

Hints and Tips

> A common mistake that people make when trying to design something completely foolproof is to underestimate the ingenuity of complete fools.
>
> —Douglas Adams

Dimensional Analysis or the Method of Dimensions

A powerful means of formulating or at least checking math equations is to use the method of dimensions. What we mean by a "dimension"—what we might call a "physics dimension"—here is a fundamental physical unit, not only a length. Most physical measurement can be expressed in terms of mass (M), length (L), and time (T). It does not matter what units you use to measure a physics dimension. A length still has dimension L, whether you measure in feet, meters, or camel-days. A mass has dimension M, whether measured in pounds, kilograms, or bushels. The different units give rise to different multiplying constants relative to each other. A quantity that is the ratio of two dimensional quantities, such as the aspect (length to width) ratio of an airplane wing, has no dimensions; we say it is "dimensionless." That applies to all things that are simply counted: for example, the number 23, of itself, is dimensionless.

You can only add up physical quantities that have the same dimensions. You can't add the weight of an orange to the diameter of an apple with any meaningful result. Trigonometric functions like sine and cosine, exponents, or logarithms *require* dimensionless quantities.

The method of dimensions applies a restriction on mathematical formulas that describe the physical world: math terms that are added or subtracted in a formula must be of the same dimensions. This follows from a moment's consideration of a formula like

$$S - \tfrac{1}{2}at^2 = 0,$$

which arises when a body is accelerated. This can be written as

$$S = \tfrac{1}{2}at^2,$$

so obviously the terms in the original formula must be of the same dimension. Change the minus sign and you will also see that added terms must be of the same dimension. In other words, in

$$S = \tfrac{1}{2}at^2 + vt$$

each variable must have the dimension L.

With a complex formula, it is sometimes worth using a spreadsheet to evaluate dimensions. For an example of using dimensional analysis "for real," using a spreadsheet, see the slimemobile project.

Occasionally you may come across problems that can usefully employ dimensions other than M, L, or T. Combinations of the fundamental M, L, and T could be used, such as V (LT^{-1}), L, and force F (MLT^{-2}). With this combination, for example, T has dimensions LV^{-1}, and M would have dimensions $FV^{-2}L$. For many ordinary problems this approach would not be helpful—but it would be effective for a problem concerning almost exclusively velocities, lengths, and forces. In a similar way, other dimensions could be used in particular problems: electric charge or current, for example, or temperature.

Data Loggers

A typical data logger is a small circuit board that converts input signals with an analog-to-digital converter (ADC) to produce digital outputs to a computer. Years ago the data logger was more typically an autonomous piece of equipment with its own recording system, either a magnetic tape or disk or an integrated-circuit memory. But these days it is easier and cheaper to purchase a small input board for a standard desktop or laptop computer. The huge memory of your computer is then available to record the data. You also thereby have the power of the computer readily at hand to analyze the data you have recorded. The easiest kind of devices to install are those that plug into a serial, parallel, or USB port, although other types that plug into a desktop machine's motherboard are available.

I often use a data logger to record the readings from an experiment automatically. Provided that you have an electrical output from an experiment, or that you can arrange for an electrical output with a suitable sensor, you can thereby avoid the tedium of writing or typing the readings. You can also gather data swiftly and without any bias. It is easy when performing manual readings to unconsciously bias your results by, for example, recording half-division values on a ruler scale less often than exact-division values: such bias adds to errors, which are often less predictable.

A useful data logger has several inputs of analog voltage in the 1 or a few volts range: most sensors or electrical inputs can be converted to lie in this range. Avoiding the use of very small signals precludes many problems, both with noise and with interference signals such as power hum. It is often helpful to have at least two channels, so that you can record *x-y* plots, or record a "baseline" or "dummy" channel as a cross-check against the channel being monitored.

Rescaling Voltages for a Data Logger

When operating a data logger, you need to "span" the inputs suitably. You should aim to scale the input so that your maximum possible input voltage from the experiment is just a little lower than the maximum voltage that the data logger will accept. In this way you can minimize the effects of noise, slight offset voltages, and other errors in the circuitry.

You can use a simple potential divider to scale down the input voltage. With a series resistor from input to data logger of value R, and a parallel resistor r across the data logger input, the resultant output voltage V_o to the data logger will be reduced relative to the input V_i according to

$$V_o = V_i r/(r + R).$$

The series resistor should not have too low a value, because it might absorb power from the input and artificially reduce the voltage measured. On the other hand, choosing too large a value (of 2M or more) might make the circuit sensitive to noise. With a small division factor, the finite resistance of the ADC circuit in the data logger will also be parallel with that of the resistor r and thus will lower the measured voltage. I often choose a value such as 100k, with a higher value if I believe the source of the voltage is of particularly high impedance.

Spreadsheet Software and Data Massaging

> To err is human.
>
> —Alexander Pope, "An Essay on Criticism," 1711

> To err is Truman.
>
> —Republican Party slogan, 1948

You often find that the results from an experiment look pretty horrible at first—full of false starts, noise, and systematic errors caused by the temperature drift of sensitive instruments. But data once logged can be "massaged" and improved using spreadsheet software.

Most data loggers store data in a simple comma-delimited file, which is highly efficient in its use of electronic storage. This format can be used as input to a spreadsheet. Spreadsheets containing the same data will be much larger, however, so once you have analyzed data and saved it in spreadsheet form, watch out. The data will occupy a much larger area of memory in the computer, which can sometimes cause problems.

Simple procedures like baseline subtraction are easily accomplished, and I regularly subtract drifting baselines, assuming that the drift is a straight line. Occasionally you will find it useful to subtract an exponential baseline, for example, in an experiment where a device is cooling down following Newton's law of cooling.

Most spreadsheets now include a whole raft of data analysis subroutines. Even more facilities are offered by specialist software packages like MathCad. Standard spreadsheets are normally sufficient, especially because they can be allied with Basic if more sophistication is required. (Spreadsheets like Excel include a version of the Basic or Visual Basic fundamental programming language behind each cell on the screen. Consult your spreadsheet manual to learn how you can access this tool.) Be sure to install all the spreadsheet's optional facilities, and then look in the toolbar for these supplemental tools. You will probably find that you have a substantial box of goodies.

Smoothing Functions

One of these tools can take an exponentially reducing weight average over a few readings before the current reading, resulting in a substantial smoothing out of jittery lines. I used this operation to clean up the readouts from the ionization sensors in the gas absorption detector project. The outputs of these sensors are intrinsically noisy because of the random production of alpha particles by the radioactive americium source, and it can be rather difficult to decipher what is happening. With a little smoothing, however, the data looked much better, and it was easy to interpret the graph in terms of the responses to different gases.

In the exponential smoothing function, try values of 0.2–0.9 for α to start with, and then judge the results on a graph. The smoothing function averages on the order of $1/(1 - \alpha)$ input readings, so with a value of 0.9, you will be averaging about ten readings. The other smoothing function generally offered is the moving average, which is a simple average of N adjacent readings.

Fourier Analysis

This tool produces the power-versus-frequency spectrum of a set of input readings. If, for example, your data follow a pure sine wave, you will see a single peak on the spectrum at the frequency of the sine wave. Some spreadsheets report the output in terms of complex numbers: you will then need to add the squares of the real and imaginary parts to get a power spectrum that you can plot. The number of data points must be a power of 2 with a typical program: 128, 256, 1,024, and so on.

Correlation Coefficients

This function produces an $N \times N$ matrix of correlation coefficients between N columns of data. Perfect correlation occurs when column I and column J are related by a linear relation like $I = KJ + K'$, where K and K' are constants. Perfect correlation produces a correlation coefficient of 1, with lesser levels of correlation yielding lower values. For example, column 1 being related to column 2 by a square relation will yield a correlation coefficient of 0.966, while a random scatter of points comes close to a coefficient value of zero. All the diagonal elements of the matrix are (obviously) 1.

Goal Seek

I don't use this option much, but it is a solution search program that finds a solution X to an algebraic equation $f(X) = Y$. In other words, to get output value Y_0, what value of X_0 do you need to put in? You have to provide a guessed value of X_0 to start the program. If the function $f(X)$ is a messy one, and if you enter a poor guesstimate as the starting value, then you may receive just one answer out of several possible answers.

Frequency Histogram

This tool is sometimes useful to make sense of data. A frequency histogram can be particularly useful if you suspect something has gone wrong with an experiment. You need to provide an input column, a column of histogram bin edges, and a space on your spreadsheet for the results. You can use bins of different widths, which can help generate a better graph of the results.

If you take multiple readings of a particular parameter from an experiment, you normally expect that nearly all the values will fall within a small range near each other, and that only a few will fall outside that small range. Most often, the frequency histogram of genuinely correct, but noisy, readings, will be a bell-shaped curve, and the normal distribution can be expressed by the equation

$$y = \exp(-\alpha[x - x_0]^2),$$

where α is the reciprocal of the width function and x_0 is the offset from zero of the bell shape.

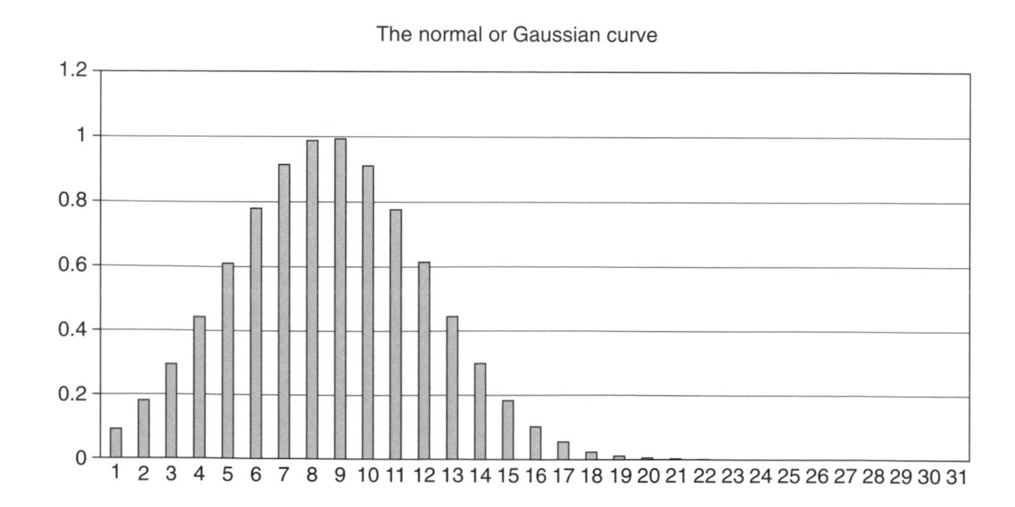

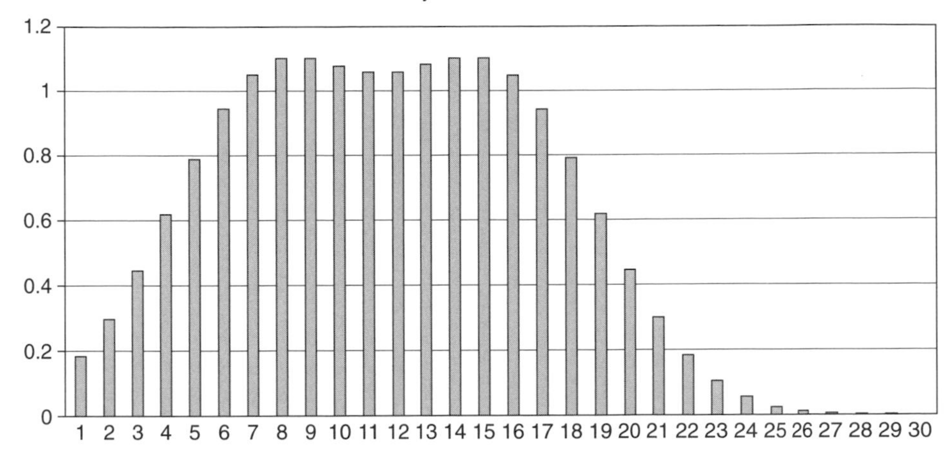

Sometimes you may find that the zero offset of the reading you are taking undergoes a shift during the experiment—maybe someone jogged your optical bench (aka the kitchen table). What you will see with a histogram analysis in this case is a combination of two sets of readings following their own bell curves with different peaks, producing a kind of Bactrian two-humped camel curve (see diagram). So if you find that your data do not fall within some kind of bell curve, be suspicious!

Hot-Melt Glue

I use this stuff all the time. It may be the modern equivalent of sealing wax in the phrase "sealing wax and string," but it is stronger and lasts longer. I also use quite a bit of cyanoacrylate (Krazy Glue). Curiously, cyano, although it often sets in a few minutes, continues to gain strength for hours on many surfaces. If you need full strength, you may find it worth leaving it to set for 24 hours. The material relies on a little moisture to set, so low ambient humidity, highly moisture-free substrates, and thick joints lead to longer cure times. (But professional suppliers like Henkel Loctite Corporation offer accelerator products; see their Web sites for more advice.)

Clamp (Retort) Stands

You often see these in chemistry labs, with a heavy iron base, a 1-m (3-foot) rod screwed into the base, a bosshead that can be fixed with a thumbscrew at any height on the vertical rod, which you can fit with a small clamping arm, with a gripper the size of a child's hand, for clasping an object. I keep several at home and find them useful for all sorts of situations when you need to hold something steady, when you need a "third hand."

When you need only a very small third hand, glue or screw an alligator clip or a clothespin onto a short piece of wood and then clamp that to the stand.

It is sometimes useful to have two or more vertical rods for an experiment. I modified one of my stands with two threaded holes in the base, on opposite sides, so that I could screw in two rods. Another way to achieve a different kind of multirod frame is to join two stands together using two spare rods and four spare bossheads. A lightweight alternative mounting is the traditional wooden rod construction toy still sold today under the Tinkertoy brand. A request to Hasbro, the Tinkertoy manufacturing company: How about a small clamp arm to go with your kit? Kids could use the arms for making robot models, and the rest of us would find Tinkertoys much more useful. Meanwhile, we can glue an alligator clip or a clothespin onto the ends of Tinkertoy rods when we want this function.

Power-Switching Relay Unit

I have found a 12-V or 6-V relay, built into an electrical extension cable unit, a useful gadget for a variety of projects. A double pole normally off relay disconnects the live and neutral wires, allowing current to flow only when low-voltage current—often only 10 or 20 mA—is supplied to the two low-voltage leads. The low-voltage control leads are brought out of a grommeted hole in the side of the extension cable end unit, with a clamp to prevent the leads from being inadvertently pulled out of the cable end unit. You can use this device whenever you need to switch a power-operated piece of equipment on and off using safe, low-voltage electrics.

Magnets

The most economical powerful permanent magnets today are still the rare-earth neodymium-iron type. They are several times more powerful than typical iron or ferrite-based magnets. A 10-mm-diameter by 5-mm-long magnet of the rare-earth kind is capable of multikilogram adhesion to a steel surface. Tiny magnets, just 3 mm in diameter and 1 mm high, which nevertheless offer a good field strength, are also available. The motors from broken radio-controlled model cars often contain powerful rare-earth samarium-cobalt magnets. Such motors are valuable parts, but if a motor is burned-out or jammed, don't discard them immediately. A little effort to dismantle and retrieve the magnets is worthwhile. Keep such motors or magnets away from the bench where you saw or grind steel parts. Metal filings will coat powerful magnets, reducing the clamping force you can get from them, and they can stop motors from turning freely. Once attached, those filings are almost impossible to remove.

Lasers

It's lucky that Noah didn't need any lasers in his ark, because lasers used to be wildly expensive. No more: laser-pointer diodes are now widely available for only a few dollars. The lifetime of some of them, particularly those from low-cost suppliers, may be short, however. I purchased one that lasted for only a few hours of continuous running. I can't really complain: after all, used normally, they would be switched on for only several seconds a few times a day, not even every day. Slightly more expensive are the lasers in laser levels sold in home-improvement stores. The lifetime should be in the thousands of hours or more if the diodes are good quality and are not run beyond their rated voltage; the cost should be no more than $25 or so. An alternative to try is one of the helium-neon laser units intended for checkout scanners in stores. Although bulky (250 mm or 10 inches long, with a fairly chunky high-voltage power supply), they have a better beam (narrower) and better color than laser diodes: a bright pink-red at a 632-nanometer wavelength (laser diodes generally feature a deep-red, 670-nanometer wavelength).

Litho Plates

Thin sheets of aluminum, around 0.2 to 0.4 mm thick, are used in huge quantities in the offset lithographic printing industry: they carry the original printing image, in photosensitive lacquer, on one side. Once used by printers they are usually discarded, although the aluminum can be recycled. The recycling value is very low, so printers will often let you have these used sheets gratis.

I find these plates quite useful. You can cut a sheet with a pair of ordinary scissors, it will not corrode too easily, and it is light and bendable. Do watch out for sharp edges, however.

Old Photocopiers

Office photocopiers are expensive machines for good reasons: they contain a treasure trove of useful mechanical, electronic, and especially electromechanical parts. So if you hear that a well-used machine is about to be discarded, hurry and claim it. Look inside for electromagnets, motors, mirrors, gearwheels, large-size capacitors, transformers, and power transistors. The purely electronic components, with the exception of some large-size items, are not, in general, worth salvaging, because they are not particularly expensive to buy new by mail order. However, a simple electromechanical component like a powerful solenoid coil is a handy gadget that is not easy to find in mail-order catalogs and will be expensive if you do find it.

English and Metric Units

Mistakes with units have caused some famous bloopers. There were spectacular mistakes even in the hallowed ranks of NASA rocket scientists with a couple of Mars space missions. I still find the old English units of measurement useful to get a feel for things, but of course the consistent metric system is much better for calculation. Here are some useful conversion factors:

1 inch	25.4 mm
1 ft	300 mm
1 mile	1.609 km
1 in^3	16.4 cm^3
1 ft^3	28.3 liters
1 oz	28.35 g
1 mph	1.609 km/h
1 hp	745 watts
Temp (F)	(1.8 × Temp (C) + 32)
–40 °F	–40 °C
32 °F	0 °C
68 °F	20 °C
212 °F	100 °C

Suggested Further Reading

Books

Alder, Henry L., and Edward B. Roessler. *Probability and Statistics.* 3rd ed. New York: W. H. Freeman, 1964.

Appleyard, Rollo. *Pioneers of Electrical Communication.* London: Macmillan, 1930. Useful short biographies of Maxwell, Ampere, Volta, Wheatstone, Hertz, Oersted, Ohm, Heaviside, Chappe, and Ronalds.

Bader, Paul, and Adam Hart-Davis. *Local Heroes: The Book of British Ingenuity.* Stroud, U.K.: Sutton, 1997.

Bagnold, Ralph A. *Physics of Blown Sand and Desert Dunes.* London: Methuen, 1941.

Balachandran, Wamadeva, Sidney A. Thompson, and S. Edward Law. "Electroclamping Forces for Controlling Bulk Particulate Flow." *Journal of Electrostatics* 37 (1996): 79–94.

———. "Metering of Bulk Materials with an Electrostatic Valve." *Transactions of the ASAE* 38 (1995): 1189–94.

———. "The Study of the Performance of an Electrostatic Valve Used for Bulk Transport of Particulate Materials." *IEEE Transactions on Industry Applications* 33 (1997): 871–78.

Balakrishnan, A. V., ed. *Advances in Communication Systems.* Vol. 1. New York: Academic Press, 1965.

Banks, Robert B. *Slicing Pizza, Racing Turtles, and Further Adventures in Applied Mathematics.* Princeton, N.J.: Princeton University Press, 1999.

Beeby, Stephen, Graham Ensell, Michael Kraft, and Neil White. *MEMS Mechanical Sensors.* Norwood, Mass.: Artech House, 2004.

Bhatia, A. B. *Ultrasonic Absorption: An Introduction to the Theory of Sound Absorption and Dispersion in Gases, Liquids, and Solids.* Oxford: Clarendon Press, 1967.

Braddick, H. J. J. *Vibrations, Waves, and Diffraction.* London: McGraw-Hill, 1965.

Braithwaite, A., and F. J. Smith. *Chromatographic Methods.* 4th ed. London: Chapman and Hall, 1985.

Bunch, Bryan, and Alexander Hellemans. *The Timetables of Science.* London: Sidgwick and Jackson, 1989.

Burke, James. *Connections.* London: Macmillan, 1978. "Connections" is also the title of Burke's monthly column in *Scientific American.*

Clark, Ronald W. *Edison: The Man Who Made the Future.* London: Macdonald and Jane's, 1977.

Coulson, J. M., J. F. Richardson, J. R. Backhurst, and J. H. Harker. *Chemical Engineering.* 4th ed. Vols. 1 and 2. Oxford: Pergamon Press, 1991.

Crawford, R. J. *Plastics Engineering.* 2nd ed. Oxford: Pergamon Press, 1989.

CRC Handbook of Chemistry and Physics. 76th ed. Boca Raton, Fla.: CRC Press, 1995.

De Bono, Edward. *Six Thinking Hats.* Boston: Little, Brown, 1985.

———. *The Use of Lateral Thinking.* New York: Harper and Row, 1970.

Doherty, Paul, and Don Rathjen. *The Exploratorium Science Snackbook Series.* New York: Wiley, 1991–96.

Downie, Neil A. *Industrial Gases.* London: Chapman and Hall, 1997.

———. *Ink Sandwiches, Electric Worms, and 37 Other Projects for Saturday Science.* Baltimore: The Johns Hopkins University Press, 2003.

———. *Vacuum Bazookas, Electric Rainbow Jelly, and 27 Other Saturday Science Projects.* Princeton N.J.: Princeton University Press, 2000.

Dummer, Geoffrey W. A. *Electronic Inventions and Discoveries.* 4th ed. Bristol, U.K.: Institute of Physics Publishing, 1997.

Eastaway, Rob, and Jeremy Wyndham. *Why Do Buses Come in Threes? The Hidden Mathematics of Everyday Life.* London: Robson Books, 1998. A fun math book.

Ehrlich, Robert. *Turning the World Inside Out.* Princeton, N.J.: Princeton University Press, 1990.

———. *Why Toast Lands Jelly Side Down.* Princeton, N.J.: Princeton University Press, 1997.

Elmore, William C., and Mark A. Heald. *Physics of Waves.* New York: McGraw-Hill, 1969.

Epstein, George. *Multiple-Valued Logic Design: An Introduction.* Bristol, U.K.: IOP Publishing, 1993.

Fauvel, John, Raymond Flood, Michael Shortland, Robin Wilson, eds. *Let Newton Be! A New Perspective on His Life and Works.* Oxford: Oxford University Press, 1988.

Feynman, Richard P. *Feynman Lectures on Computation.* London: Penguin Books, 1999.

Feynman, Richard P., Robert B. Leighton, and Matthew Sands. *The Feynman Lectures on Physics.* Reading, Mass.: Addison-Wesley, 1963. Three volumes from one of the grand masters of physics, now dated in some ways but still excellent.

Fleming, J. A. *Electric Wave Telegraphy and Telephony.* London: Longmans, 1916.

Ford, R. A. *Homemade Lightning: Creative Experiments in Electricity.* 3rd ed. New York: McGraw-Hill, 2002.

Garwin, Laura, and Tim Lincoln. *A Century of Nature: Twenty-one Discoveries that Changed Science and the World.* Chicago: University of Chicago Press, 2003.

Greene, David. *Light and Dark: An Exploration in Science, Nature, Art and Technology.* Bristol, U.K.: Institute of Physics Publishing, 2003. A well-written, fascinating account of light-related phenomena from sunlight to moonbeams via optical fibers by a professor of semiconductor science.

Gregory, Richard L. *Eye and Brain: The Psychology of Seeing.* 5th ed. Princeton, N.J.: Princeton University Press, 1997.

Griskey, Richard G. *Polymer Process Engineering.* New York: Chapman and Hall, 1995.

Grosvenor, E. S., and Morgan Wesson. *Alexander Graham Bell.* New York: Harry N. Abrams, 1997.

Hall, Christopher. *Polymer Materials: An Introduction for Technologists and Scientists.* 2nd ed. New York: Wiley, 1989.

Highfield, Roger, and Peter Coveney. *The Arrow of Time.* London: Flamingo, 1991.

Hill, K. M., and J. Kakalios. "Reversible Axial Segregation of Binary Mixtures of Granular Materials." *Physical Review E* 49 (1994): 3610–13.

Hill, Winfield, and Paul Horowitz. *The Art of Electronics.* Cambridge: Cambridge University Press, 1989. This *magnum opus* was the bible of electronics students for years.

Hopkins, George M. *Experimental Science: Elementary, Practical and Experimental Physics.* 1902. Reprint, Bradley, Ill.: Lindsay Publications, 1997. Two volumes with lots of Victorian everyday science and science demonstrations.

Inwood, Stephen. *The Man Who Knew Too Much: The Strange and Inventive Life of Robert Hooke, 1635–1703.* London: Macmillan, 2002. Superb account of this rather obscure early figure in science (remembered today primarily for Hooke's law of springs), who made many discoveries key to the modern world.

James, Peter, and Nick Thorpe. *Ancient Inventions.* London: Michael O'Mara Books, 1994.

Jargodski, Christopher P., and Franklin Potter. *Mad about Physics: Braintwisters, Paradoxes, and Curiosities.* New York: Wiley, 2001.

Jewkes, John, David Sawers, and Richard Stillerman. *The Sources of Invention.* London: Macmillan, 1958.

Kirk-Othmer Concise Encyclopedia of Chemical Technology. 4th ed. New York: Wiley, 1999.

Kittel, Charles. *Introduction to Solid State Physics.* 5th ed. New York: Wiley, 1976.

Koerner, Thomas W. *The Pleasures of Counting.* Cambridge: Cambridge University Press, 1997. Excellent anecdotes about the application of math to the real worlds of industry and war.

Laithwaite, E. R. *Propulsion without Wheels.* 2nd ed. London: English Universities Press, 1971.

Landes, David S. *Revolution in Time: Clocks and the Making of the Modern World.* Cambridge, Mass.: Harvard University Press, 2000. This superb book covers clocks and their historical impact on society.

Lavoisier, A. L. *Elements of Chemistry.* France, 1789. For a more recent edition, see Hutchins, Robert M. *Great Books of the Western World.* Vol. 45, *Lavoisier, Fourier, Faraday.* Trans. Robert Kerr. Chicago: William Benton/Encyclopaedia Britannica, 1952.

Lindley, David. *Degrees Kelvin: The Genius and Tragedy of William Thomson.* London: Aurum Press, 2004. Biography of Lord Kelvin, the Victorian pioneer of global communications technology, who also contributed hugely to the science of thermodynamics and developed the modern magnetic compass.

Lovelock, James. *Gaia: A New Look at Life on Earth.* 1979. Reprint with corrections and a new preface. Oxford: Oxford University Press, 2000.

Loxton, R., and P. Pope, eds. *Instrumentation: A Reader.* London: Chapman and Hall, 1990. A collection of superb articles on different kinds of sensors and instruments by the experts who make and work with them.

Michels, W. J., C. E. Wilson, and J. P. Sadler. *Kinematics and Dynamics of Machinery.* New York: HarperCollins, 1983.

Moroney, M. J. *Facts from Figures.* London: Pelican/Penguin Books, 1951.

Needham, Joseph, with Wang Ling and Kenneth Girdwood Robinson. *Science and Civilisation in China.* Vol. 4, *Physics and Physical Technology.* Cambridge: Cambridge University Press, 1962–65. I have found vol. 4 (in two parts) of this work the most interesting. In total, *Science and Civilisation in China* runs to dozens of heavy volumes. The magnum opus of Professor Needham, it covers superbly the development of science and technology in the almost completely isolated Chinese empires of the last three thousand years.

O'Hare, Mick, ed. *The Last Word: Questions and Answers from the Popular Column on Everyday Science.* Oxford: Oxford University Press, 1998.

———. *The Last Word 2: More Questions and Answers on Everyday Science.* Oxford: Oxford University Press, 2000.

Pawle, Gerald. *The Secret War, 1939–1945.* London: Harrap, 1956.

Rabinowicz, Ernest. *Friction and Wear of Materials.* 2nd ed. New York: Wiley, 1995.

Rosenberg, H. M. *The Solid State.* Oxford: Clarendon Press, 1975.

Sekuler, Robert, and Randolph Blake. *Perception.* 2nd ed. New York: McGraw-Hill, 1990.

Semmens, P. W. B., and A. J. Goldfinch. *How Steam Locomotives Really Work.* Oxford: Oxford University Press, 2000.

Sharpe, Carill, ed. *Kempe's Engineers' Yearbook.* Tonbridge, U.K.: Miller-Freeman, 1996.

Simons, Martin. *Model Aircraft Aerodynamics.* 2nd ed. Hemel Hempstead, Hertfordshire, U.K.: Argus Books, 1987.

Singer, Charles, E. J. Holmyard, A. R. Hall, and Trevor I. Williams, eds. *A History of Technology.* Oxford: Oxford University Press, 1954–58. This is a masterful work of five huge volumes (subsequently extended to seven), running from early prehistory to 1900. In its thousands of pages are thousands of lost technologies, which make fascinating reading and might have practical uses today.

Singh, Simon. *The Code Book: The Science of Secrecy from Ancient Egypt to Quantum Cryptography.* London: Fourth Estate, 1999.

Smiles, Samuel. *Lives of the Engineers: The Steam Engine, Boulton and Watt.* London: John Murray, 1878. A venerable but still fascinating account not only of the birth of steam power but also of eighteenth-century capitalism.

Sokolnikoff, I. S., and E. S. Sokolnikoff. *Higher Mathematics for Engineers and Physicists,* chap. 11. London: McGraw-Hill, 1941.

Solymar, Laszlo. *Getting the Message: A History of Communications.* Oxford: Oxford University Press, 1999.

Soucek, Ludvik. *The Story of Communications.* London: Mills and Boon, 1969. A curious but amusing illustrated book translated from the Czech.

Standage, Tom. *The Victorian Internet: The Remarkable Story of the Telegraph.* New York: Walker Publishing, 1998.

Steeds, William. *Mechanism and the Kinematics of Machines.* London: Longmans, Green, 1940.

Stephenson, Geoffrey. *Mathematical Methods for Science Students.* 2nd ed. Harlow, U.K.: Longman, 1973.

Stewart, Ian. *Game, Set, and Math: Enigmas and Conundrums.* Oxford: Basil Blackwell, 1989.

Stremler, Ferrel G. *Introduction to Communication Systems.* 2nd ed. Reading, Mass.: Addison-Wesley, 1982.

Tables of Physical and Chemical Constants. Originally compiled by G. W. C. Kaye and T. H. Laby, now prepared under the direction of an editorial committee. 16th ed. Harlow, U.K.: Longman, 1995.

Thomson, J. J. *Conduction of Electricity through Gases.* Cambridge: Cambridge University Press, 1906.

Thomson, N., ed. *Thinking Like a Physicist.* Bristol, U.K.: Adam Hilger, 1987.

Uglow, Jenny. *The Lunar Men: The Friends Who Made the Future, 1730–1810.* London: Faber, 2002.

Usher, Abbott Payson. *A History of Mechanical Inventions.* 1934. Reprint, New York: Dover, 1988.

Van Doren, Carl, ed. *Benjamin Franklin's Autobiographical Writings.* London: Cresset Press, 1946. Contains interesting writings on science, as well as samples of Franklin's everyday and political wit and wisdom.

Vines, A. E., and N. Rees. *Plant and Animal Biology.* 4th ed. London: Pitman, 1972.

Vogel, Steven. *Life in Moving Fluids: The Physical Biology of Flow.* 2nd ed. Princeton, N.J.: Princeton University Press, 1994.

Wang, T. T., J. M. Herbert, and A. M. Glass, eds. *The Applications of Ferroelectric Polymers.* Glasgow: Blackie, 1988.

Web Sites and Periodicals

For free access to the eleventh (1911) edition of the *Encyclopaedia Britannica,* visit www.1911encyclopedia.org. I also possess and frequently use the 1952 print edition of the *Encyclopaedia Britannica,* obtained very inexpensively at a garage sale. These older editions of this treasure trove contain, if anything, more science and technology than today's editions, often explaining them in more detail. And of course they record forgotten technologies with historical interest.

The U.S. Patent Office, the European Patent Office, and, to a lesser extent, the British Patent Office are gold mines, with details of more than 30 million inventions and discoveries, mostly in English or with English abstracts, from around 1600 to the present day. All three offer searchable Web sites: www.uspto.gov/; http://ep.espacenet.com/; and www.patent.gov.uk/. The European Patent Office's Web site is probably the best.

Scientific American magazine has been around since 1850, and I find the issues before 1930 or so to be particularly interesting today. *New Scientist* is a highly readable, United Kingdom–based science weekly. *The Proceedings of the Royal Institution of Great Britain* (London, 1799 on) publishes tidied-up transcripts of lectures aimed at people with a scientific background but no specialist expertise in a field; this journal often provides masterful overviews of a subject. *The Proceedings of the Royal Society* (London, 1661 on) often includes more maverick contributions than most scientific journals. *Eureka Innovative Engineering Design,* a United Kingdom–based magazine, is basically an advertising vehicle for engineering suppliers, but it has short articles on genuinely new technologies.

Late Victorian technical and scientific journals offer a fascinating glimpse into a world that may look rather modern in photographs but was, underneath, vastly different from today's world. One such periodical is *Technics* (1904–5; it may have continued after 1905, but those are the only issues I have found). Another publication worth exploring is *The English Mechanic and World of Science,* a weekly newspaper published from 1865 to 1926.

Index